Spectroscopy and Machine Learning for Water Quality Analysis

IOP Series in Spectroscopic Methods and Applications

Series Editor
Ashutosh Kumar Shukla, Ewing Christian College, University of Allahabad, Prayagraj, India

About the Series

The *IOP Series in Spectroscopic Methods and Applications* aims to develop a library of key texts encompassing the broad range of spectroscopic techniques, instrumentation and fields of applications found in modern science. Some examples of the spectroscopic techniques incorporated include, IR, UV/VIS, Raman, EPR, NMR, fluorescence spectroscopy, force spectroscopy, hyperspectral imaging, neutron scattering, Mössbauer spectroscopy, photoemission spectroscopy and all the various specialist techniques within these broad domains and more. Microscopic characterization techniques may also be included where they are used in conjunction with spectroscopic methods for sample analysis.

The series contains two broad types of approach. Those addressing a particular field of application and reviews the numerous relevant spectroscopic methods applicable to the field, and those that focus on a specific spectroscopic method which will permit a greater in-depth review of the theory and instrumentation technology.

Spectroscopy and Machine Learning for Water Quality Analysis

Edited by
Ashutosh Kumar Shukla
Ewing Christian College, University of Allahabad, Prayagraj, India

IOP Publishing, Bristol, UK

ISBN 978-0-7503-3047-3 (ebook)
ISBN 978-0-7503-3045-9 (print)
ISBN 978-0-7503-3048-0 (myPrint)
ISBN 978-0-7503-3046-6 (mobi)

DOI 10.1088/978-0-7503-3047-3

Version: 20210401

IOP ebooks

British Library Cataloguing-in-Publication Data: A catalogue record for this book is available from the British Library.

Published by IOP Publishing, wholly owned by The Institute of Physics, London

IOP Publishing, Temple Circus, Temple Way, Bristol, BS1 6HG, UK

US Office: IOP Publishing, Inc., 190 North Independence Mall West, Suite 601, Philadelphia, PA 19106, USA

To my wife.

Contents

Preface

This collection intends to highlight the applications of spectroscopic and other techniques for water quality analysis and monitoring. There are five chapters in this volume by experts. Chapter 1 by Daniel Cozzolino presents the origin of IR spectra, sampling techniques and recent applications of Fourier transform mid infrared (FTMIR) spectroscopy in water analysis. Chapter 2 by Cong Wang and Daoliang Li highlights the need of new monitoring tools to overcome the shortcomings of traditional ones. They have reviewed the applications of fluorescence spectrometry in water quality monitoring and potential applications of fluorescence sensors. Chapter 3 by Bambang Kuswandi presents the applications of paper-based optical sensors in water quality analysis and monitoring. Exciting avenues to enrich water quality analysis experience by integration of such sensors with Artificial Intelligence and Machine Learning have been presented in this chapter. Chapter 4 by Naumih M Noah and Peter M Ndangili reviews the role of nanocomposites for water treatment. Synthesis, characterization and applications of polymeric and non-polymeric nanocomposites have been discussed in detail in this chapter. Chapter 5 by Sachchidanand Soaham Gupta and Rajeev Singh presents basics and applications of Machine Learning and Artificial Neural Network models for water quality assessment.

I sincerely thank John Navas, Senior Commissioning Manager, IOP publishing for giving me an opportunity to present this book to readers. I wish to thank Sarah Armstrong, Editorial Assistant, IOP Publishing for extending all the support during this project. My heartfelt thanks to the expert authors for contributing to this volume.

Ashutosh Kumar Shukla
Prayagraj, India

Editor biography

Shukla Ashutosh Kumar

Ashutosh Kumar Shukla is Associate Professor of Physics at Ewing Christian College, Prayagraj, UP, India. His current research interests include Electron Spin Resonance studies of doped double perovskites. Dr Shukla received his BSc, MSc (Physics), and DPhil from University of Allahabad. With two decades of Physics teaching, he has to his credit peer reviewed journal publications, conference presentations, invited lectures, review articles, text book, and several edited volumes by reputed publishers.

List of contributors

Professor Daniel Cozzolino

Appointed as Principal Research Fellow/Associate Professor QAAFI—University of Queensland (January 2020). Dr Cozzolino was an Associate Professor in Food Chemistry in the School of Science at RMIT University (Melbourne, VIC) (2018–2020), Head of Agriculture and Associate Professor in Agriculture at CQUniversity (Rockhampton, QLD) (2015–2019), Senior Research Fellow—Barley Breeding Program at The University of Adelaide (Adelaide, SA) (2013–2015) and Team Leader of the Rapid Analytical Group at The Australian Wine Research Institute (Adelaide, SA) (2002–2012). He has published more than 300 peer-reviewed articles (*h* index 59, 10 500 citations Google Scholar). His research focusses on the developments of chemometric and spectroscopic methods (e.g., UV–vis, NIR, MIR) applied in agriculture (e.g., crops, animal products), environment, and food (e.g., composition, process, authentication, and provenance). Recent research projects have spanned the authentication, fraud and origin of foods, the effect of climate change in food quality (e.g., grain), process analytics, in field and high-throughput techniques, monitoring plastics in the food chain, identification of markers in biological tissues (e.g., cancer cells). In 2013, he was presented with the Tomas Hirschfeld Award by the *International Council for NIR Spectroscopy* for his outstanding contributions in the field of NIR spectroscopy.

Professor Daoliang Li

Daoliang Li is full professor of College of Information and Electrical Engineering and the director of the National Innovation Center for Digital Fishery, China Agricultural University. He received the PhD degree from the College of Engineering, China Agricultural University, his principal research interest is ICTs in aquaculture and agriculture, especially for information processing, smart sensors and smart control system in fish farming and aquaponics. He is the editor-in-chief of *International Journal of Information Processing in Agriculture* (https://www.journals.elsevier.com/information-processing-in-agriculture/) and the Chair of the Work Group for Advanced Information Processing in Agriculture, International Federation for Information Processing. He is also the deputy director for the Expert Consultant Committee for National Agricultural and Rural informatization, Ministry of Agriculture and Rural Development. He was the chairman of 1st to 13th International Conference on Computer and Computing Technologies in Agriculture (www.iccta.cn). He coordinated more than 100 international and national research projects, such as FP6, FP7, Horizon2020 and has published more than 300 international journals papers, eight books. He also has owned 31 Chinese patents.

Dr Cong Wang

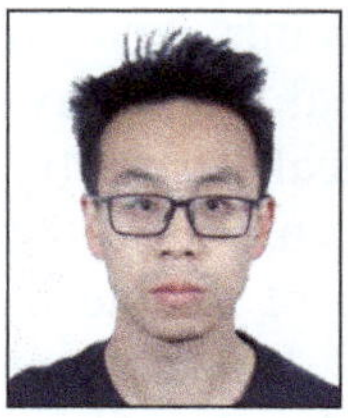

Cong Wang is a postdoctoral of College of Engineering, China Agricultural University. He received a PhD degree in 2020 from the College of Engineering, China Agricultural University. His principal research interest is automatic measurement and control techniques in fish farming and aquaponics.

Professor B Kuswandi

Professor Bambang Kuswandi received his PhD in Instrumentation and Analytical Science from University of Manchester Institute of Science and Technology (UMIST), Manchester, UK in 1999 under supervisor Professor Dr R Narayanaswamy. He also did his MSc degree (1996) in the same department, at UMIST, UK and undergraduate in Chemical Education (1993) IKIP Malang (now UM, State University of Malang), Indonesia.

He is currently head of Chemo and Biosensors Group, Professor in Analytical Chemistry (2009). He was formerly Head of School of Pharmacy (2003–2009) and Dean of Faculty of Pharmacy, University of Jember, Indonesia (2009–2013). Professor Kuswandi was a Visiting Research Scientist at Department of Chemistry, Hacettepe, Ankara Turkey (1997), Department of Instrumentation and Analytical Science, UMIST, UK (2000), Department of Chemistry, University of Indonesia (2001); Department of Chemistry, University of Florence, Italy (2004); Laboratory of Supramolecular Chemistry and Technology, MESA+ Research Institute for Nanotechnology, University of Twente, the Netherlands (2006–2007) and Department of Chemistry and Biochemistry, Georgia Institute of Technology, Atlanta, Georgia, USA (2008–2009). Visiting Professor at School of Chemical Sciences & Food Technology, Faculty of Science & Technology, UKM Malaysia (2011–2012). Currently, he is a Visiting Professor at IChemTech, Faculty of Science & Technology & Fellow at IHRAM (Institute of Halal Research & Management), Universiti Sains Islam Malaysia (2013–2017), and Honorary visiting at School of Chemistry, the University of Melbourne, Melbourne, Australia (2018). He got *h* index 20 and 1418 citations (Scopus) for his over 50 papers published in refereed journals. His current research interests are development of chemical sensor and biosensor for applications in environmental, food, pharmacy, clinical and health monitoring, including smart packaging and sensors for natural product pharmacy, optical and electrochemical sensors, molecular recognition, host–guest chemistry, ligand design and synthesis, the use of polymers and other novel/smart materials as solid support, reagent and indicator in sensors development, lab on a-chip and microfluidics devices, development of wearable sensing system, and halal science and technology.

Dr Naumih Noah

Dr Naumih Noah is an analytical/bioanalytical chemist and an Assistant Professor at the United States International University-Africa. Her research is on analytical/bioanalytical chemistry focusing on the development and validation of biosensors for the detection of infectious diseases and the integration of nanotechnology and health care in developing countries. Her current research is on the development of Nano-kits for rapid and sensitive diagnostic of Schistosomiasis, green synthesis of nanoparticles as well as fabrication of nanofilters for water filtration. She is a trained chemical safety and security expert and is also interested in green chemistry and waste management. She has authored over 20 peer-reviewed journal articles and three book chapters on Nanotechnology.

Dr Peter Ndangili

Dr Peter Ndangili is a senior lecturer of analytical/physical chemistry at the Technical University of Kenya. His research is on the fabrication of smart sensors and biosensors using electroactive nanomaterials; which include nanopolymers, polymeric nanocomposites and chalcogenide quantum dots. His focus in sensor/biosensor application is in disease diagnostics, detection of water pollutants, cosmetic dentistry, and transgenic detection in genetically modified food crops. His current research is on the development of Electrochemical Interdigitated Quantum dot, waveguide sensors for detection of pathogenic *Escherichia coli* O157:H7 and Vibrio cholera in vegetable salads and fruit juices as well as fabrication of nanofilters for water filtration. He is a trained chemical security and safety expert and a certified reviewer of the American Chemical Society journals. He has authored 17 articles in peer reviewed journals and two book chapters.

Sachchidanand Soaham Gupta

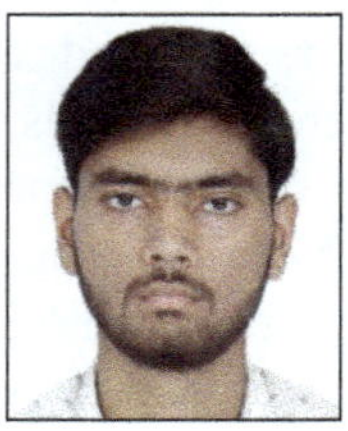

Soaham is currently pursuing his graduate degree in Applied Chemistry from Amity University Noida. He has completed his bachelor's in Industrial Chemistry from Atma Ram Sanatan Dharma College, University of Delhi. During his undergraduate degree, he had demonstrated a keen interest in research. He wishes to pursue his career as a researcher and pursue a PhD.

Dr Rajeev Singh

Rajeev Singh received his PhD degree in inorganic/organometallic chemistry from the Department of Chemistry, University of Delhi, India. He joined as Chemistry Faculty, Atma Ram Sanatan Dharma College, New Delhi in 2009. Singh did his postdoctoral studies as TWAS-CNPq Postdoc Fellow at Universidade Federal do Ceara, Brazil. Currently he is also the Incharge of Centre for Innovation and Entrepreneurial Leadership, Technology Business Incubator. He is the recipient of Dr Karkhanavala Award 2006 by ITAS-BARC. He has completed many sponsored research projects. His major research areas are coordination chemistry, organometallic complexes, material chemistry, and hybrid materials. He has delivered lectures at various international and national conferences and has nearly 30 international publications.

Dr Pratibha Chaudhary

Dr. Pratibha Chaudhary is currently working as Assistant Professor in Department of Chemistry, Maitreyi College, University of Delhi. Her field of specialization is organic chemistry.

IOP Publishing

Spectroscopy and Machine Learning for Water Quality Analysis

Ashutosh Kumar Shukla

Chapter 1

FTIR spectroscopy and water quality

Daniel Cozzolino

The advantages of Fourier transform mid infrared (FTMIR) spectroscopy have been known for more than a century where applications of this technique can be found in the field of chemistry, for example, in the analysis of pure compounds, polymers. More complex molecules have many bonds where vibrations can also be conjugated leading to two possible modes of vibration: stretching and bending. Despite these potential issues, absorption frequencies might be used to identify specific chemical groups, and this capability has traditionally been the main role of FTMIR spectroscopy for the routine analysis of different type of samples. This chapter provides an overview of the origin of the spectra in the infrared region (IR), sampling techniques, and provides examples of recent applications of FTMIR spectroscopy in water analysis.

1.1 Introduction

Applications of Fourier transform mid infrared (FTMIR) spectroscopy have been developed over the past decades owing to their many advantages of this technique when compared with the use of other traditional routine methods of analysis. Some of these advantages include the non-destruction of the sample, minimal or no sample preparation and does not involve the use of any hazardous chemicals during the analysis (Ciurczak 2001, McClure 1994, 2001, 2003, Barton 2002). In addition, one of the inherent characteristics of this technique, the time required during the analysis of a given sample, determine that a few seconds or minutes are needed for data collection (e.g., scanning) allowing for the analysis (e.g., prediction using calibrations) of several properties that can be estimated from a single measurement.

Some of these advantages of FTMIR spectroscopy have been known for more than a century where applications of this technique can be found in the field of chemistry, for example, in the analysis of pure compounds, polymers, among others. This technique has evolved from the development of the interferometer by Michelson in the 1880s (Ciurczak 2001, McClure 1994, 2001, 2003, Barton 2002)

doi:10.1088/978-0-7503-3047-3ch1

and the proposed transformation by Lord Rayleigh in relation to the interference patterns produced by the interferometer that could be converted into a spectrum using Fourier transformation (Ciurczak 2001, McClure 1994, 2001, 2003, Barton 2002). After the second World Word dispersive instruments had been widely used and replaced by modern FTIR instruments in the late 1960s (McClure 2001, Barton 2002).

This chapter provides an overview of the origin of the spectra in the infrared region (IR), sampling techniques, and provides examples of recent applications of FTMIR spectroscopy in water analysis.

1.2 The origin of the spectra

Near infrared (NIR) and mid infrared (MIR) spectroscopy have been considered as non-destructive and non-invasive measurements where the major advantage of NIR over MIR spectroscopy, is that NIR light can penetrate much further into a sample than MIR radiation (Murray and Williams 2001, Ciurczak 2001, McClure 2001, Barton 2002, Stuart 2004).

Considering reflectance spectra in the visible (VIS) (400–780 nm) and NIR (780–2500 nm) regions of the electromagnetic spectrum, they are the result of interactions between the radiating energy and the different chemical bonds in the molecules of organic compounds (Murray and Williams 2001, Ciurczak 2001, McClure 2001, Barton 2002, Stuart 2004). For example, in the VIS region, the high energy of the radiation causes transitions of electrons between molecular orbits with different energy levels while with lower radiation energy, corresponding to longer wavelengths, the absorption of energy occurs due to vibrations in molecular bonds (Murray and Williams 2001, McClure 1994, 2003). In contrast, in the MIR region absorption bands arise from the functional groups of such properties, the so called fundamental frequencies (Murray and Williams 2001, McClure 1994, 2003).

The absorptions in the NIR region are due to the presence of overtones and combinations of fundamental vibrations that are originated in the MIR region (Murray and Williams 2001, McClure 1994, 2003). The energy quanta absorbed are bond specific, but are also affected by the chemical matrix and other factors such as type of functional group, neighboring molecules and hydrogen bonds (Osborne *et al* 1993, Murray and Williams 2001, McClure 2003). These properties allow for the identification of a range of molecules which may contain the same type of bonds (Murray and Williams 2001, McClure 1994, 2003). The same molecule can give rise to several overtones and combination bands over the whole NIR region with decreasing intensity and increasing overtone order (Murray and Williams 2001, McClure 1994, 2003). Because of this, the NIR region is characterized by few, broad, overlapping absorption features (Murray and Williams 2001).

On the other hand, the spectral 'signatures' in the MIR are the result from the fundamental stretching, bending, and rotating vibrations of the sample molecules, whilst NIR spectra result from complex overtones and high frequency combinations at shorter wavelengths (Weyer 1985, Osborne *et al* 1993, Murray and Williams 2001, Wetzel 2001, McClure 2003). Spectral intensities in the MIR frequencies are often sharper and better resolved than in the NIR domain (Weyer 1985, Osborne *et al* 1993, Murray and Williams 2001, Wetzel 2001, McClure 2003). Therefore, different bonds can be identified and analyzed, such as the higher overtones of the O–H (oxygen–hydrogen), N–H (nitrogen–hydrogen), C–H (carbon–hydrogen), and S–H (sulfur–hydrogen) (Weyer 1985, Osborne *et al* 1993, Murray and Williams 2001, Wetzel 2001, McClure 2003). Although, bands from the MIR region are still observed in the NIR they are characterized as much weaker compared with those present in the MIR region (fundamental frequencies) (Weyer 1985, Osborne *et al* 1993, Murray and Williams 2001, Wetzel 2001, McClure 2003). In addition to the existence of combination bands (e.g., CO stretch and NH derived from proteins), the NIR region is also characterized by strongly overlapping bands (Weyer 1985, Osborne *et al* 1993, Murray and Williams 2001, Wetzel 2001, McClure 2003).

A major disadvantage of this characteristic overlap and complexity in the NIR spectra has been the difficulty of quantification and interpretation of data from NIR spectra (Weyer 1985, Osborne *et al* 1993, Murray and Williams 2001, Wetzel 2001, McClure 2003). On the other hand, the broad overlapping bands can diminish the need for using a large number of wavelengths during the analysis (Weyer 1985, Osborne *et al* 1993, Murray and Williams 2001, Wetzel 2001, McClure 2003).

Chemical bonds present in the organic matrix of most of the samples vibrate at specific frequencies, which are determined by the mass of the constituent atoms, the shape of the molecule, the stiffness of the bonds, and the periods of the associated vibrational coupling (Weyer 1985, Osborne *et al* 1993, Murray and Williams 2001, Wetzel 2001, McClure 2003). A specific vibrational bond absorb in the IR spectral region where diatomic molecules have only one bond that may stretch (e.g., the distance between two atoms may increase or decrease) (Weyer 1985, Osborne *et al* 1993, Murray and Williams 2001, Wetzel 2001, McClure 2003).

More complex molecules have many bonds where vibrations can also be conjugated leading to two possible modes of vibration namely stretching and bending (Chalmers and Griffiths 2002, McClure 1994, 2003). Despite these potential issues, the absorption frequencies might be used to identify specific chemical groups and this capability has traditionally been the main role of Fourier transform (FT) MIR (FTMIR) spectroscopy (Chalmers and Griffiths 2002, McClure 1994, 2003). Molecules absorb in the infrared (IR) due to the vibrational movements of their chemical bonds, for example, bending, stretching, rocking, wagging or scissoring (Chalmers and Griffiths 2002, McClure 1994, 2003). As the movements occur at specific energy levels, chemical bonds absorb IR at specific wavenumbers or wavelengths correlating to their different energy levels. Consequently, IR spectroscopy is the measure of this absorption, resulting in a spectrum with peaks representing the chemical bonds present in the given sample (Chalmers and Griffiths 2002, McClure 1994, 2003).

The MIR region (4000–400 cm^{-1}) is particularly important for biological applications and contains the so called biochemical and chemical fingerprint region (1800–900 cm^{-1}) (Chalmers and Griffiths 2002, McClure 1994, 2003, Karoui *et al* 2010, Li-Chan 2010). This region is known to be absorbed by a number of biochemical and chemical compounds such as Amide I (~1650 cm^{-1}), Amide II

(~1550 cm^{-1}), Amide III (~1260 cm^{-1}), carbohydrates (~1155 cm^{-1}), glycogen (~1030 cm^{-1}), lipids (~1750 cm^{-1}), asymmetric phosphate stretching vibrations (PO_2^-; ~1225 cm^{-1}), symmetric phosphate stretching vibrations (PO_2^-; ~1080 cm^{-1}), and protein phosphorylation (~970 cm^{-1}) (Chalmers and Griffiths 2002, McClure 1994, 2003, Karoui *et al* 2010, Li-Chan 2010). In particular the MIR range between 1500 and 400 cm^{-1} the so called fingerprint region can be found (Chalmers and Griffiths 2002, McClure 1994, 2003, Karoui *et al* 2010, Li-Chan 2010). Absorptions in the fingerprint region are mainly caused by bending and skeletal vibrations, which are particularly sensitive to large wavenumber shifts, thereby minimizing against unambiguous identification of specific functional groups (Chalmers and Griffiths 2002, McClure 1994, 2003, Karoui *et al* 2010, Li-Chan 2010). Analysis of such fingerprints is the basis of many applications of MIR spectroscopy (Chalmers and Griffiths 2002, McClure 1994, 2003, Karoui *et al* 2010, Li-Chan 2010).

1.3 Hardware/instrumentation

A typical instrument consists of a radiation source, a wavenumber or frequency selection device such as a monochromator, a sample holder, a photoelectric detector that measures the intensity of detected light and converts it to electrical signals, and a computer system that acquires and processes spectral data (Bell 1972, Beckman *et al* 1977, Coates 1997, Ferraro 1999, Christy *et al* 2001). Traditional dispersive grating MIR instrumentation had similar optical systems to those present in NIR instruments, where an IR source passes through a sample, then through a monochromator that contains optical devices such as a prism or grating (Griffiths and de Haseth 1986, Griffiths *et al* 1972, McCarthy and Kemeny 2001, Subramanian and Rodrigez-Saona 2009). The light is guided through the prism or grating and separated into a spectrum of different wavelengths. An exit slit is employed to allow light at specific wavelengths to be segmented into readily measurable components and pass through to reach the detector. In situations where a grating or prism is used, these can be rotated to allow the light at different wavelengths to pass through the exit slit (Griffiths and de Haseth 1986, Griffiths *et al* 1972, McCarthy and Kemeny 2001, Smith 1996, Subramanian and Rodrigez-Saona 2009). Both FTMIR and FTNIR spectrometers operate by applying Fourier transform to interferograms obtained from a Michelson interferometer with a movable mirror (Griffiths and de Haseth 1986, Griffiths *et al* 1972, McCarthy and Kemeny 2001, Smith 1996, Subramanian and Rodrigez-Saona 2009). Modern MIR instruments are FTMIR instruments based on the Michelson interferometer (Griffiths and de Haseth 1986, Griffiths *et al* 1972, McCarthy and Kemeny 2001, Smith 1996, Subramanian and Rodrigez-Saona 2009). Detectors are optical devices designed to measure the intensity of the IR light that strikes them by converting radiation energy into electrical signals.

1.4 Sampling techniques

The application of FTMIR spectroscopy is of special interest due to the presence of sharp and specific absorption bands for constituents (Griffiths and de Haseth 1986,

Griffiths *et al* 1972, McCarthy and Kemeny 2001, Smith 1996). More recently the development of sampling accessories attached to a wide range of FTMIR spectrophotometers, such as attenuated total reflectance (ATR) cells, has led to major improvements in routine IR analysis, by simplifying sample handling and avoiding measurement problems often found using transmission cells (Subramanian and Rodrigez-Saona 2009, Karoui *et al* 2010, Li-Chan 2010, Milosevic 2012). Several sampling methods exist where each of the sampling techniques has its own advantages and limitations related with their end use application (Subramanian and Rodrigez-Saona 2009, Karoui *et al* 2010, Li-Chan 2010, Milosevic 2012).

The simplest sampling technique is transmittance. This sampling technique involves passing the IR radiation directly through the sample and detecting on the other side. It provides spectra with high signal-to-noise ratio (SNR) and is relatively cost effective. This sampling technique is suited for the analysis of solid, liquid, and gaseous samples (Subramanian and Rodrigez-Saona 2009, Karoui *et al* 2010, Li-Chan 2010, Milosevic 2012). However, transmission technique is limited by the sample thickness. For example, samples within the range 1–20 μg are suited for transmission analysis (Smith 1996). This sampling technique requires a very careful and time consuming sample preparation (Subramanian and Rodrigez-Saona 2009, Milosevic 2012). It is well known that IR transparent materials such as KB are used as a substrate for the sample during analysis (Subramanian and Rodrigez-Saona 2009, Karoui *et al* 2010, Li-Chan 2010, Milosevic 2012). Reflectance is the reverse of transmittance in which IR light reflected back from the sample is measured. Reflectance is not influenced by the sample thickness where the SNR is typically lower than that observed for the case of transmittance (Subramanian and Rodrigez-Saona 2009, Karoui *et al* 2010, Li-Chan 2010, Milosevic 2012). Furthermore, the depth of penetration into the sample is not exactly known and the surface of the sample influences the spectra more than the interior. These disadvantages hidden by the simplicity and speed of this technique. Based on the type of reflectance from the sample, reflectance technique also includes specular reflectance, diffused diffuse reflectance, and ATR (Subramanian and Rodrigez-Saona 2009, Karoui *et al* 2010, Li-Chan 2010, Milosevic 2012). ATR has been one of the most commonly used sampling techniques in recent times (Subramanian and Rodrigez-Saona 2009, Karoui *et al* 2010, Li-Chan 2010, Milosevic 2012). When an IR beam travels from a medium of high refractive index (e.g., zinc selenide crystal) to a medium of low refractive index (sample), some amount of the light is reflected back into the low refractive index medium (Subramanian and Rodrigez-Saona 2009, Karoui *et al* 2010, Li-Chan 2010, Milosevic 2012). At a particular angle of incidence, almost all of the light waves are reflected back. This phenomenon is called total internal reflection (Milosevic 2012). In this condition, some amount of the light energy escapes the crystal and extends a small distance (0.1–5 μm) beyond the surface in the form of waves. This invisible wave is called evanescent wave. The intensity of the reflected light reduces at this point. This phenomenon is called ATR (Subramanian and Rodrigez-Saona 2009, Milosevic 2012). When the sample is applied on the crystal some amount of the IR radiation penetrating beyond the crystal is absorbed by the sample. This absorbance is translated into the IR spectrum of the sample

(Subramanian and Rodrigez-Saona 2009, Karoui *et al* 2010, Li-Chan 2010, Milosevic 2012). Diffuse reflectance occurs when the IR beam is reflected back from the sample surface in random direction. It involves both absorption and scattering. The scattering is due to the rough sample surface, which is the category that most foods belong to. Diffuse reflectance is well suited for solid and powder materials (Subramanian and Rodrigez-Saona 2009, Karoui *et al* 2010, Li-Chan 2010, Milosevic 2012). A diffuse reflectance accessory employs two flat mirrors to direct the light and one concave focusing mirror exactly above the sample cup to concentrate the IR beam on to the sample (Subramanian and Rodrigez-Saona 2009, Karoui *et al* 2010, Li-Chan 2010, Milosevic 2012). Specular reflectance is a type of reflectance technique that occurs when the angle of incidence of the IR radiation incident on the sample is equal to the angle at which it is reflected back. Reflectance on a mirror surface is a typical example of specular reflectance (Subramanian and Rodrigez-Saona 2009, Karoui *et al* 2010, Li-Chan 2010, Milosevic 2012).

1.5 Examples of applications of MIR spectroscopy in water analysis

Despite the well-known advantages and characteristics of FTMIR spectroscopy, not many applications can be found in the literature on applications or uses of this technique to analyze water and water components. The following section will provide an overview of some of the recent applications of this technique.

The IR spectra of water in aqueous solutions (e.g., hydrogen bonded systems) have distinctive bands that are originated from several overlapping components (Max and Chapados 2015). Several attempts to identify the origin of these bonds were reported in the literature (Max and Chapados 2015). The use of derivatives and in particular the use of Gaussian, Lorentzian (Cauchy), and Gauss–Lorentz has been reported to understand the origin of the overlapping bonds associated with water (Max and Chapados 2015). Liquid propanol, liquid acetonitrile, and aqueous acetone have been used to model and identify the origin of these bands (Max and Chapados 2015).

The use of ATR–MIR spectroscopy was also evaluated with mixtures of glycerol/water solutions in order to gain a better understanding of the strong hydrogen bonding of glycerol (Kataoka *et al* 2011). These authors evaluated that the OH stretching band after eliminating the contribution of glycerol OH in the glycerol/water solutions was decomposed using three Gaussian components (Kataoka *et al* 2011). These authors also suggested that with increasing glycerol concentrations up to 50% (v/v), the decrease the absorption at 3428 cm^{-1} and the increase of the 3562 cm^{-1} suggesting a breaking of H bonds among water molecules while the absorbance at 3242 cm^{-1} was unchanged (Kataoka *et al* 2011). The authors also discussed that water molecules surrounding glycerol molecules are retained by strong H bonding between H atoms of water and O atoms in C–O of glycerol when aqueous solutions containing glycerol are introduced in human skin (Kataoka *et al* 2011).

Recent studies by Mecozzi *et al* (2009a, 2009b) and Mecozzi and Pietroletti (2016) reported the spectral characteristics of synthetic mucilage samples using two dimensional (2D) IR spectroscopy (Mecozzi and Pietroletti 2016, Mecozzi *et al* 2009a, 2009b). These authors have reported this for *Asparagopsis armata* mucilage samples and the mixture of brown algae mucilage samples. The authors reported a strong auto-peak on the diagonal was observed at 3350 cm^{-1} in synchronous 2D IR spectra, which mainly comes from the absorption of OH group, indicating that hydroxyl plays a key role in the formation of stable mucilage (Mecozzi *et al* 2009a, 2009b, Mecozzi and Pietroletti 2016). In addition, the authors also reported weak auto peaks at 1125 cm^{-1} (carbohydrate) and 1650 cm^{-1} (protein), which indicated that both carbohydrate sand proteins can assist the formation of mucilage (Mecozzi *et al* 2009a, 2009b, Mecozzi and Pietroletti 2016). It was pointed out by the authors that the OH group plays an important role in the formation of mucilage, and the other groups play an auxiliary role (Mecozzi *et al* 2009a, 2009b, Mecozzi and Pietroletti 2016). In another study, the same authors further tested the technologies of 2D wavelength–wavelength and sample–sample correlation spectroscopy to investigate the roles played by carbohydrates, proteins and lipids in the process of aggregation of natural marine organic matter (OM) (Mecozzi *et al* 2009a, 2009b, Mecozzi and Pietroletti 2016). The results reported by these authors showed that carbohydrates and proteins are involved in the aggregation process (Mecozzi *et al* 2009a, 2009b, Mecozzi and Pietroletti 2016). The authors concluded that the use of 2D infrared correlation spectroscopy was also a powerful tool for the elucidation of the mechanisms involved in the aggregation of natural marine OM (Mecozzi *et al* 2009a, 2009b, Mecozzi and Pietroletti 2016).

A direct method for real-time monitoring of ammonia (NH_3) concentration in fish ponds was reported by Goldshleger *et al* (2018). The method was based on monitoring the spectral signatures of pure ammonia in both the MIR and far-infrared range (Goldshleger *et al* 2018). The authors reported a linear relationship between ammonia concentration during the evaporation process and the detected infrared signals (Goldshleger *et al* 2018). Ammonia concentrations in the order of 1 ppm were detected with excellent agreement ($<\pm 5\%$) with conventional methods (Goldshleger *et al* 2018). The authors concluded that the advantages of the method were the direct measurement of ammonia rather than the indirect calculations and measurements currently used, providing with better accuracy, and real-time operation with minimal maintenance of the analytical system (Goldshleger *et al* 2018).

The mixing properties of glycerol–water and diglycerol–water solutions were measured using ATR–MIR spectroscopy (Habuka *et al* 2020). These authors reported that absorbance spectra corrected for 1 μm thickness were subtracted by pure polyols for obtaining water spectra, and by pure water for polyol spectra (Habuka *et al* 2020). These authors reported that both asymmetric and symmetric CH_2 stretching vibration bands (around 2940, 2885 cm^{-1}) shifted about 10 cm^{-1} to lower wavenumbers with increasing polyol concentrations, especially at higher concentrations (Habuka *et al* 2020). Shifts of C–O–H rocking bands (around 1335 cm^{-1}) with increasing polyol concentrations are slightly larger for diglycerol–water (10 > 6 cm^{-1}) than glycerol–water solutions where C–O stretching bands of CHOH groups (1125 and 1112 cm^{-1}) shift slightly but in opposite sides for glycerol and diglycerol at highest polyol concentrations (90–100 wt%) (Habuka *et al* 2020). These shifts of CH_2 stretching,

COH rocking, and CO stretching of CHOH at higher polyol concentrations suggest interactions of outer CH_2 with inner CHOH groups of surrounding polyols (Habuka *et al* 2020). The normalized band area changes with polyol concentrations could be fitted by quadratic polynomials possibly due to mixtures of different interactions between water–water, polyol–water, and polyol–polyol molecules (Habuka *et al* 2020). The OH stretching band for diglycerol 90 wt% shows three humps indicating at least three OH components: long, medium, and short H bond water molecules (Habuka *et al* 2020). Short H bond water molecules are the major component possibly between inner CHOH and outer side CH_2OH groups, while the long H component might loosely bind to outer CH_2OH groups (Habuka *et al* 2020).

Two-dimensional correlation (2DCOS) analyses were applied to the time-dependent ATR–FTIR spectra of aqueous solutions of α- and β-D-glucose undergoing such mutarotation conversion (Noda 2018). The use of 2DCOS analysis reveals that the increase and decrease in the IR intensities after the dissolution of α- or β-D-glucose are not fully synchronized, suggesting the mutarotation of D-glucose in water is not a simple binary conversion process but a multi-step reaction involving an intermediate species with a finite and observable concentration level and lifetime (Noda 2018). The 2DCDS analysis of the time-dependent ATR–FTIR spectra clearly demonstrated the presence of intermediate species contributing to the band positions overlapped close to bands for α- and β-D-glucose (Noda 2018). The fact that band positions identified for the intermediate species for α- to β-D-glucose conversion are the same for the reverse reaction suggests that they arise from the same species, most likely the open-ring structure produced by the hydrolysis (Noda 2018).

The contamination of drinking water with arsenic was investigated using different techniques such as ATR–FTIR, Raman spectroscopy, and x-ray diffraction (XRD) (Puente-Urbina and Montero-Campos 2017). In this study, the co-precipitation method was applied as reference method to modify kaolin, bentonite, diatomite, and silica in order to prepare magnetic composite nanoparticles having support-magnetite weight ratios of 2:1 (Puente-Urbina and Montero-Campos 2017). The method has the capacity to remove up to 80% and 70% for initial concentrations of 25 and 50 $\mu g\ L^{-1}$, respectively, representing a convenient remover for processes developed in subsequent stages or in continuous flow (Puente-Urbina and Montero-Campos 2017). These authors reported that the nanoparticles showed absorbances at 3417 cm^{-1} (stretching of O–H bonds) and at 1600 cm^{-1} (H–O–H bending in water molecules) (Puente-Urbina and Montero-Campos 2017). The authors have also reported that stabilized nanoparticles (nanoparticles coated with carboxylic acids) presented differences region corresponding with to O–H bond stretching, at 2921 cm^{-1} and 2852 cm^{-1} (symmetric and asymmetric methylene stretching of the coating agents), and at 1519 cm^{-1} and 1428 cm^{-1} (symmetric and asymmetric stretching vibrations of carboxyl groups) (Puente-Urbina and Montero-Campos 2017). The presence of these frequencies suggested the surface modification of the nanoparticles by the carboxylate groups of the carboxylic acids (Puente-Urbina and Montero-Campos 2017). The authors indicated that use of ATR–MIR spectroscopy enabled providing information in order to characterize the nanoparticles (Puente-Urbina and Montero-Campos 2017).

Ion pairs in aqueous solution were quantitatively and simultaneously determined using ATR–FTIR spectroscopy (Rauh and Mizaikoff 2016). Seven salts present in natural seawater, such as NaCl, KCl, NaBr, KBr, $MgCl_2$, $CaCl_2$, and Na_2SO_4 were evaluated (Rauh and Mizaikoff 2016). The authors used multivariate data analysis to discriminate and assign the spectral information originated from each salt in calibration mixtures, using mixtures of all constituents in different concentrations (Rauh and Mizaikoff 2016). The multivariate method proposed by the authors was able to discriminate between NaCl/KCl, NaBr/KBr, $MgCl_2$, $CaCl_2$, and Na_2SO_4 with sodium and potassium chloride, and bromide being treated as sum parameters, respectively (Rauh and Mizaikoff 2016). An additional multivariate model was able to discriminate between NaCl, KCl, NaBr, and KBr including their simultaneous quantification (Rauh and Mizaikoff 2016). The authors validated the models using a real seawater sample where $MgCl_2$ could be correctly quantified at 0.5% and 0.05% (w/v) (Rauh and Mizaikoff 2016).

Understanding colloid composition has been frequently a key issue in order to predict contaminant transport in natural waters (McGuire and Herman 2015). Most of the current methods to collect and identify colloid composition chemically alter the colloid before analysis and consequently fail to identify the mineralogy and organic components of the colloid under analysis (McGuire and Herman 2015). The authors evaluated the use of ATR–FTIR spectroscopy to identify colloids including organic material in concentrated suspensions (McGuire and Herman 2015). The authors concluded that ATR–FTIR spectroscopy allows for direct analysis of concentrated suspensions rather than requiring drying of the samples which alters the composition of the sample in the colloidal phase (e.g., precipitation of carbonates in samples from karst waters) (McGuire and Herman 2015). The authors concluded that the ability of ATR–FTIR to monitor variations in colloidal composition was demonstrated through the examination of colloids under two different flow conditions namely a karst aquifer and the West Branch of the Susquehanna River in Central Pennsylvania (McGuire and Herman 2015).

1.6 Concluding remarks

The application of MIR spectroscopy allows the collection of the so called spectral 'signatures' that result from the fundamental stretching, bending, and rotating vibrations of different molecules that constitute a given sample. These signatures provide the means to quantify different organic molecules and the interaction of inorganic molecules with water and other particles in different matrices. In addition, the development of sampling techniques such as diffuse reflectance, ATR, transmission and more recently FT micro-spectroscopy increase the number of applications and the ability of MIR spectroscopy to analyze different samples.

Adapting and using advanced technologies (e.g., vibrational spectroscopy and chemometrics) are a promising way to monitor different biological systems towards sustainability. These new tools will allow the industry and researchers alike to move forward and develop novel applications that can address best management practices.

However, various hurdles still hinder the growth and development of these technologies. The reluctance of the industry to accept the incorporation of vibrational spectroscopy with new statistical tools such as chemometric techniques as routine analytical methods. Furthermore, most of the current courses and training programs in analysis and technology still focus on the so-called classical approach, which do not incorporate several aspects related to new technologies, sensors and programming in their curricula. The same can apply to research and other aspects of informal training and extension.

References

Barton F E 2002 Theory and principles of near infrared spectroscopy *Spectrosc. Eur.* **14** 12–8

Beckman A O, Gallaway W S, Kaye W and Ulrich W F 1977 History of spectrophotometry at Beckman instruments, Inc *Anal. Chem.* **49** 280A–300A

Bell R J 1972 *Introductory Fourier Transform Spectroscopy* (New York: Academic)

Chalmers J M and Griffiths P R 2002 *Handbook of Vibrational Spectroscopy* (New York: Wiley)

Ciurczak E W 2001 Principles of NIR spectroscopy *Handbook of Near-Infrared Analysis* 2nd edn ed D A Burns and E W Ciurczak (Boca Raton, FL: CRC Press) pp 8–9

Coates J 1997 Vibrational spectroscopy: instrumentation for infrared and Raman spectroscopy *Analytical Instrumentation Handbook* 2nd edn ed G W Ewing (New York: Marcel Dekker)

Christy A A, Ozaki Y and Gregoriou V G 2001 *Modern Fourier Transform Infrared Spectroscopy* (New York: Elsevier)

Ferraro J R 1999 History of Fourier transform-infrared spectroscopy *Spectroscopy* **14** 28–40

Griffiths P R and de Haseth J A 1986 *Fourier Transform Infrared Spectrometry* (New York: Wiley)

Griffiths P R, Foskett C T, Curbelo R and Dunn S T 1972 Rapid-scan infrared Fourier transform spectroscopy *Appl. Spectrosc. Rev.* **6** 31–78

Goldshleger N, Grinberg A, Harpaz S, Shulzinger A and Abramovich A 2018 Real-time advanced spectroscopic monitoring of Ammonia concentration in water *Aquacult. Eng.* **83** 103–8

Habuka A, Yamada T and Nakashima S 2020 Interactions of glycerol, diglycerol, and water studied using attenuated total reflection infrared spectroscopy *Appl. Spectrosc.* **74** 767–79

Karoui R, Downey G and Blecker C 2010 Mid-infrared spectroscopy coupled with chemometrics: a tool for the analysis of intact food systems and the exploration of their molecular structure-quality relationships – a review *Chem. Rev.* **110** 6144–68

Kataoka Y, Kitadai N, Hisatomi O and Nakashima S 2011 Nature of hydrogen bonding of water molecules in aqueous solutions of glycerol by attenuated total reflection (ATR) infrared spectroscopy *Appl. Spectrosc.* **65** 436

Li-Chan E C Y 2010 Introduction to vibrational spectroscopy *Applications of Vibrational Spectroscopy in Food Science* ed E Li-Chan, P R Griffiths and J M Chalmers (New York: Wiley)

Max J-J and Chapados C 2015 Determination of spectroscopic band shapes using second derivatives, part I: theory *Appl. Spectrosc.* **69** 348

McCarthy W J and Kemeny G J 2001 Fourier transform spectrometers in the near-infrared *Handbook of Near-Infrared Analysis* ed D A Burns and E W Ciurczak (New York: Taylor & Francis) pp 71–90

McClure W F 1994 Near-infrared spectroscopy *Spectroscopic Techniques for Food Analysis* ed R H Wilson (New York: VCH Publishers) pp 13–52

McClure W F 2001 Near-infrared instrumentation *Near-infrared Technology in the Agricultural and Food Industries* 2nd edn ed P C Williams and K H Norris (St. Paul, MN: AACC) pp 96–8

McClure W F 2003 204 years of near infrared technology: 1800–2003 *J. Near Infrared Spectrosc.* **11** 487–518

McGuire M M and Herman E K 2015 A novel ATR–FTIR technique for identifying colloid composition in natural waters *Hydrol. Process.* **29** 1314–23

Mecozzi M and Pietroletti M 2016 Chemical composition and surfactant characteristics of marine foams investigated by means of UV–vis, FTIR and FTNIR spectroscopy *Environ. Sci. Pollut. Res.* **23** 22418–32

Mecozzi M, Pietroletti M, Gallo V and Conti M 2009a Formation of incubated marine mucilages investigated by FTIR and UV–vis spectroscopy and supported by two-dimensional correlation analysis *Mar. Chem.* **116** 18–35

Mecozzi M, Pietrantonio E and Pietroletti M 2009b The roles of carbohydrates, proteins and lipids in the process of aggregation of natural marine organic matter investigated by means of 2D correlation spectroscopy applied to infrared spectra *Spectrochim. Acta* **71** 1877–84

Milosevic M 2012 *Internal Reflection and ATR Spectroscopy* (New York: Wiley)

Murray I and Williams P C 2001 Chemical principals of near-infrared technology *Near Infrared Technology in the Agricultural and Food Industries* 2nd edn ed P C Williams and K H Norris (St. Paul, MN: AACC) pp 22–3

Noda I 2018 Two-dimensional correlation and codistribution spectroscopy (2DCOS and 2DCDS) analyses of time-dependent ATR IR spectra of D-glucose anomers undergoing mutarotation process in water *Spectrochim. Acta A: Mol. Biomol. Spectrosc.* **197** 4–9

Osborne G B, Fearn T and Hindle P H 1993 *Practical NIR Spectroscopy with Applications in Food and Beverage Analysis* (New York: Longman Scientific & Technical)

Puente-Urbina A and Montero-Campos V 2017 Porous materials modified with Fe_3O_4 nanoparticles for arsenic removal in drinking water *Water Air Soil Pollut.* **228** 374

Rauh F and Mizaikoff B 2016 Simultaneous quantification of ion pairs in water via infrared attenuated total reflection spectroscopy *Anal. Methods* **8** 2164

Smith B C 1996 *Fundamental of Fourier Transform Infrared Spectroscopy* (Boca Raton, FL: CRC Press)

Subramanian A and Rodrigez-Saona L 2009 Fourier transform infrared (FTIR) spectroscopy *Infrared Spectroscopy for Food Quality Analysis and Control* ed D W Sun (Oxford: Academic) pp 146–74

Stuart B 2004 Modern infrared spectroscopy *Analytical Chemistry by Open Learning Series* ed D J Ando (Chichester: Wiley) pp 180

Weyer L G 1985 Near-infrared spectroscopy of organic substances *Appl. Spectrosc. Rev.* **21** 1–43

Wetzel D L 2001 Contemporary near-infrared instrumentation *Near-infrared Technology in the Agricultural and Food Industries* 2nd edn ed P C Williams and K H Norris (St. Paul, MN: AACC) pp 129–44

IOP Publishing

Spectroscopy and Machine Learning for Water Quality Analysis

Ashutosh Kumar Shukla

Chapter 2

Fluorescence spectroscopy and applications in water quality monitoring

Cong Wang and Daoliang Li

With the rapid development of industry, more and more pollutants are accumulated in cities, and these pollutants flow into rivers, lakes, and oceans, which not only threaten the environment, but also threaten human drinking water and the growth of aquatic life. It is very important to monitor the change of water quality quickly and accurately, find out pollution events and take corresponding measures. At present, environmental monitoring departments usually use conventional chemical methods to analyze water quality. However, with the increasing complexity of the pollution situation and the continuous change of people's requirements for pollution monitoring indicators, such methods expose many shortcomings. For example, the above-mentioned tests are very time-consuming and complicated, and cannot monitor the real-time change of water quality in time. Therefore, environmental monitoring departments urgently need new monitoring means to overcome the shortcomings of traditional detection methods. As one of the methods of water quality analysis, fluorescence spectrometry has the characteristics of fast, high sensitivity, high selectivity, fewer samples and no damage to the sample structure. Molecular fluorescence spectroscopy provides a potential application for water quality monitoring of oceans, large lakes, and rivers. In this chapter, the research progress of fluorescence analysis in the measurement of chlorophyll, colored dissolved organic matter, dissolved oxygen, pH, ammonia nitrogen, nitrite, nitrate, and heavy metal in water in recent years is reviewed. The design principle, detection mechanism and the relationship between structure and chemical sensor of fluorescence probe are mainly analyzed, which provides guidance for the application of fluorescence sensors in the measurement of different water quality indices.

doi:10.1088/978-0-7503-3047-3ch2

2.1 Introduction

Water quality monitoring is a process of monitoring and measuring the types of pollutants in water, the concentrations and changing trends of various pollutants, and evaluating the water quality. The monitoring range is very wide, including uncontaminated and polluted natural water (river, lake, sea, and groundwater) and various forms of industrial drainage. The main monitoring items can be divided into two categories: one is the comprehensive indicators reflecting the water quality, such as temperature, chroma, turbidity, pH value, conductivity, suspended solids, dissolved oxygen, chemical oxygen demand, and biochemical oxygen demand; the other is some toxic substances, such as phenol, cyanogen, arsenic, lead, chromium, cadmium, mercury, and organic pesticides. In order to objectively evaluate the water quality of rivers and oceans, in addition to the above monitoring items, it is sometimes necessary to measure the flow rate.

At present, water quality monitoring technology is mainly physical and chemical monitoring technology, including chemical method, electrochemical method, atomic absorption spectrophotometry, ion selective electrode method, ion chromatography, gas chromatography, plasma emission spectrometry (ICP-AES), etc. In recent years, biological monitoring, remote sensing monitoring technology has also been applied to water quality monitoring. Among these, fluorescence analysis is a method of qualitative or quantitative analysis based on the principle that some substances will produce fluorescence which can reflect the characteristics of the substance after being irradiated by ultraviolet. Compared with the electrochemical method and mass spectrometry, fluorescence analysis equipment is simple, sensitive and selective, and is widely used in water quality analysis. Based on the analysis of the principle and method of fluorescence analysis, this chapter discusses the application of fluorescence analysis in water quality analysis.

2.2 Principle and method of fluorescence analysis

2.2.1 Principle

At room temperature, most molecules are in the ground state. When the molecules are irradiated by a certain wavelength of light, they will transition from the ground state to the excited state, and then most molecules will quickly fall back to the ground state, release photons through radiation decay, and produce fluorescence. The energy of transition from the ground state to the excited state of different elements is not the same, so the fluorescence color and intensity are not the same when the radiation decays back to the ground state. Fluorescence analysis is a qualitative or quantitative analysis method based on the fluorescence produced by different elements and the type and content of elements according to the fluorescence wavelength and intensity. Fluorescence analysis needs to be based on emission spectrum and excitation spectrum. The excitation spectrum is the spectrum of the excitation wavelength corresponding to the fluorescence intensity obtained by changing the excitation wavelength under the condition of fixed fluorescence emission wavelength. Emission spectrum is to obtain the spectrum of fluorescence intensity and corresponding emission wavelength by changing the emission wavelength under the condition of fixed excitation wavelength and intensity. Fluorescence analysis can not only detect the substances that emit fluorescence, but also detect some inorganic and organic substances that do not emit fluorescence by using fluorescent reagent.

2.2.2 Methods

At present, the classical fluorescence analysis methods can be divided into three categories: fluorescence generation method, catalytic fluorescence method and fluorescence quenching method. Fluorescence generation method is a direct method for the determination of fluorescence generation. It uses organic fluorescence reagents for determination and analysis. This kind of method has high sensitivity but weak specific selectivity. Catalytic fluorimetry is a method of fluorescence analysis by catalytic reaction. The object of catalytic fluorimetry is not the substance itself, but other substances amplified by catalytic reaction, with high sensitivity and high detection limit.

Fluorescence quenching refers to all phenomena that lead to the decrease of fluorescence intensity and lifetime of a specific substance. There are mainly two ways of fluorescence quenching: dynamic quenching and static quenching. Dynamic quenching is a process of interaction between quencher and excited molecules of fluorescent substances, such as energy transfer process or electron transfer process caused by collision. The reason for dynamic quenching may also be that the fluorescent substance generates instantaneous ground state complexes under the action of quenching agent, and these ground state complexes may not emit fluorescence, or the fluorescence emission characteristics are different from the original fluorescence characteristics. Static quenching is the complex reaction between the fluorescent substance in the ground state and the quencher to produce a non-fluorescent complex. Due to the paramagnetism of oxygen molecules, the transition between systems is promoted, and the excited singlet state of fluorescent molecules is transformed into triplet state, which leads to the internal energy conversion, and the energy is lost in the form of heat rather than fluorescence. Dynamic quenching can shorten the fluorescence lifetime due to the energy transfer, while static quenching can only reduce the fluorescence intensity by combining the quencher with the fluorescent substances to form a non-fluorescent substance, and the fluorescence lifetime does not change.

Fluorescence spectrometry is a fast, simple and inexpensive method to determine the concentration of an analyte in solution based on its fluorescent properties. It can be used for relatively simple analyses, where the type of compound to be analyzed ('analyte') is known, to do a quantitative analysis to determine the concentration of the analytes. Fluorescence is used mainly for measuring compounds in solution. In fluorescence spectroscopy, a beam with a wavelength varying between 180 and ~800 nm passes through a solution in a cuvette. We then measure—from an angle—the light that is emitted by the sample (see figure 2.1). In fluorescence spectrometry both an excitation spectrum (the light that is absorbed by the sample) and/or an emission

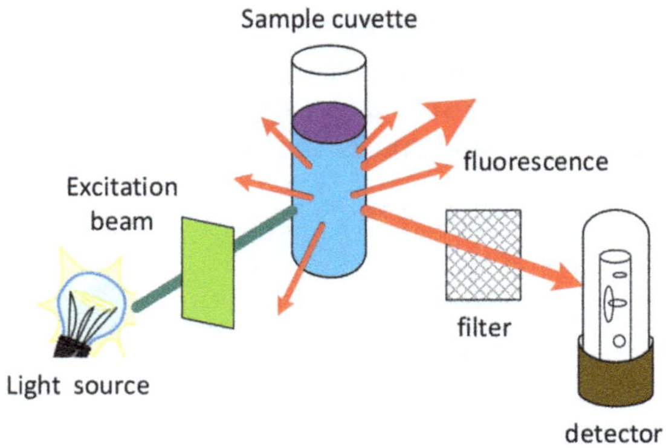

Figure 2.1. Fluorimeter scheme. The light source is in a 90° angle with the detector. The sample is located at the intersection of the two beam paths.

spectrum (the light emitted by the sample) can be measured. The concentration of the analyte is directly proportional with the intensity of the emission.

There are several parameters influencing the intensity and shape of the spectra. When recording an emission spectrum the intensity is dependent on the excitation wavelength, the concentration of the analyte solvent, the path length of the cuvette and the self-absorption of the sample.

2.3 Application of fluorescence analysis in water quality monitoring

2.3.1 Chlorophyll and CDOM

Phytoplankton mass growth is the main phenomenon of water eutrophication, in which chlorophyll a is the type of chlorophyll contained in all phytoplankton categories. It is not only an important indicator of the division of water nutritional status, but also can be used to characterize the standing crop of phytoplankton. Therefore, chlorophyll a plays a key role in the evaluation of water eutrophication. The on-line monitoring of chlorophyll a is usually carried out by photoluminescence analysis.

As interest in the aquatic cycle of organic carbon (OC) has increased, the deployment of *in situ* optical sensors to measure CDOM (colored dissolved organic matter) fluorescence as a proxy for OC concentration has become more common. CDOM sensors typically use UV light (~350 nm) to excite the emission of blue light (~450 nm) from certain organic fluorophores, allowing investigators to distinguish CDOM from more commonly measured phytoplankton pigments (Watras *et al* 2011).

At present, the chlorophyll-a sensor based on fluorescence method has made some progress. A chlorophyll fluorescence sensor was designed by Lamb *et al* (2012). Three kinds of LEDs (435 nm, 470 nm, and 457 nm) were used as the fluorescence excitation light source of chlorophyll-a solution, chlorophyll-b solution and chlorophyll-a, -b mixed solution, respectively. A photodiode integrated with a preamplifier was used as the detector. The dynamic range of the sensor was 0–5 μmol L^{-1}. Yin *et al* (2018) designed a tunable pulse induced fluorescence sensor based on fluorescence dynamics, using a 465 nm LED array as excitation light source and photomultiplier tube as detector. The intermediate parameters of chlorophyll-a fluorescence induction process were detected by fast strong pulse light and slow weak pulse light, and the electron conversion rate was inversed, then the photosynthetic reaction rate was measured. Blockstein and Yadid-Pecht (2014) designed a lensless portable chlorophyll-a sensor, which uses 465 nm LED as light source, CMOS as detector, and camera lens coated with Bismarck brown Y (BBY) as long pass filter. The dynamic range of the sensor is 0–1 μmol L^{-1} by running image processing algorithm on FPGA platform. Rao *et al* (2011) designed a chlorophyll and turbidity sensing system, using 470 nm high power LED and three color LED (470 nm, 525 nm, 660 nm) as light sources, respectively, for fluorescence detection and absorbance detection, and using a temperature sensor to compensate the temperature drift of the system. Hu *et al* (2013) proposed a weak optical signal measurement method based on pseudo-random sequence under bandwidth constraint, which optimizes the relationship between M-sequence length and measurement accuracy under limited bandwidth, and improves the suppression ability of the measurement system to narrow-band interference. Based on this method, the *in situ* chlorophyll-a sensor, *in situ* turbidity sensor and *in situ* CDOM sensor are designed. The M-sequence modulation with code length of 31 bits is used. The sensitivities are 0.0103 μg L^{-1} (chlorophyll-a calibration), 0.0078 FTU (formazin calibration) and 0.137 μg L^{-1} (quinine sulfate calibration), respectively. Table 2.1 lists the performance parameters of some commercial chlorophyll fluorometers. LOD is the abbreviation of limit of detection.

MicroFlu-CDOM is a low-cost miniaturized submergeable fluorometer for high precision and selective colored dissolved organic matter (CDOM) fluorescence measurements in the environment. The combination of a miniaturized design with long-term stability makes it suitable for yellow substance monitoring applications in lakes and rivers as well as for all applications in the field of water quality and wastewater monitoring. Internal reference measurement of the emitted light compensates aging and temperature dependences of the high-efficient LEDs, used for fluorescence excitation.

Sharikova *et al* developed a laser-induced fluorescence (LIF) system to detect and continuously observe in real time the levels of CDOM or dissolved organic compounds (DOCs) in water from various sources, such as tap water and reverse osmosis processed water (Sharikova and Killinger 2010). The portable LIF system had two interchangeable microchip Nd:YAG lasers, operating at 266 nm and 355 nm, as UV sources,

Table 2.1. Commercial chlorophyll fluorometer.

Name	Light source	λ_{ex} (nm)/ λ_{em} (nm)	Measurement mode	Range (μg L^{-1})	Limit of detection (μg L^{-1})
Trios MicroFlu	LEDs	470/685	*In situ*	0–200	0.1
Turner	LEDs	470/685	*In situ*	0–500	0.025
Seapoint	LEDs	470/685	*In situ*	0–150	0.1
YSI 6025	LEDs	470/685	*In situ*	0–400	0.1

Table 2.2. Commercial CDOM fluorometer.

Name	Light source	λ_{ex} (nm)/ λ_{em} (nm)	Measurement mode	Range (μg L^{-1})	Limit of detection (μg L^{-1})
MicroCDOM	LEDs	370/460	*In situ*	0–200	0.2
Turner	LEDs	370/460	*In situ*	0–500	0.4
Seapoint	LEDs	370/440	*In situ*	0–1500	0.1
YSI EXO	LEDs	370/460	*In situ*	0–300	0.07

and fluorescence was measured over the range of 260–680 nm. The fluorescence was collected at 90° to the laser beam. The signal-to-noise ratio (SNR) achieved with 266 nm laser excitation was about 10 times greater than that obtained with 265 nm LED. The laser pulse energy, however, was about 100 times greater than the LED's. Similarly, SNR obtained with 355 nm laser excitation was about 5 times greater than SNR achieved with 355 nm LED, while their pulse energies differed by a factor of 20. Table 2.2 lists the performance parameters of some commercial CDOM fluorometers.

2.3.2 Dissolved oxygen

The determination of oxygen concentration is important in many areas of industry, medicine and the environment. The amount of oxygen dissolved in water is an indication of the quality of the water and careful control of oxygen levels is important in wastewater management and in fermentation processes. On-line monitoring of dissolved oxygen plays an important role in sustainable intensive aquaculture, because it can ensure a fish farm of high yield, efficient, safe, and healthy by perceived aquatic products' living environment in the breeding process.

In the process of stimulated emission fluorescence, dynamic quenching and static quenching exist at the same time. Therefore, there are two ways to detect fluorescence quenching: detecting fluorescence lifetime and detecting fluorescence intensity. According to the residual fluorescence intensity after quenching of dissolved oxygen in the sample solution, the weaker the fluorescence intensity is, the greater the quenching degree, which means the higher the concentration of dissolved oxygen is, in contrast, the lower the concentration of dissolved oxygen. The dissolved oxygen concentration can be calculated directly according to the Stern–Volmer equations. The method is simple, but it has higher requirements for the stability of light source and the filtering of stray light. The oxygen quenching process is described by the Stern–Volmer equations:

$$\frac{I_0}{I} = 1 + K_{SV}pO_2 \quad (2.1)$$

$$\frac{\tau_0}{\tau} = 1 + K_{SV}pO_2 \quad (2.2)$$

$$K_{SV} = k\tau_0 \quad (2.3)$$

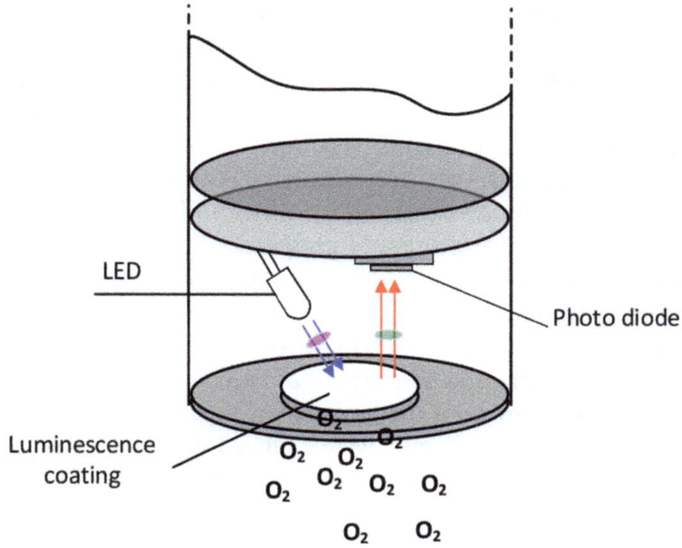

Figure 2.2. Internal structure of dissolved oxygen sensor by fluorescence method.

where I and τ are, respectively, the fluorescence intensity and excited state lifetime of the fluorophore, the subscript 0 denotes the absence of oxygen, K_{SV} the Stern–Volmer constant, k the diffusion-dependent bimolecular quenching constant, and pO_2 the oxygen partial pressure. Figure 2.2 represents the typical internal structure of an optical dissolved oxygen sensor.

The optical oxygen sensors based on intensity quenching of the fluorescence have many drawbacks. These include susceptibility to lightsource and detector drift, to changes in optical path, and drift due to degradation or leaching of the dye. These effects scan be minimized by operating the sensor in the time domain instead of the intensity domain. The lifetime, τ, is an intrinsic property of the fluorophore which, unlike intensity, is virtually independent of external perturbations. The lifetime is quenched (reduced) in the presence of oxygen and this quenching is described in equation (2.2). If the excitation signal is sinusoidally modulated, the dye fluorescence is also modulated but is time delayed or phase shifted relative to the excitation signal. The relationship between the lifetime, τ, and the corresponding phase shift, ϕ, for a single exponential decay is:

$$\tau = \frac{\tan \phi}{2\pi f} \tag{2.4}$$

where f is the modulation frequency. A disadvantage of phase fluorometry is that the SNR decreases with increasing modulation frequency, and since the phase sensitivity increases with modulation frequency, an optimal frequency has to be selected.

McDonagh *et al* developed a high performance dissolved oxygen probe, which was based on phase fluorometric detection of the quenched fluorescence of an oxygen-sensitive ruthenium complex ($[Ru^{II}$–Tris(4,7-diphenyl-1,10-phenanthroline)$]^{2+}$,

$(Ru–(Ph_2phen)_3^{2+}))$ (McDonagh *et al* 2001). The LED excitation (λ_{ex} = 500 nm) and photodiode detection (λ_{em} = 610 nm) were employed in a dipstick probe configuration, with the oxygen-sensitive film coated on a disposable PMMA disc, which in turn was designed to guide light into the film by total internal reflection. A dual LED referencing system and careful selection of optoelectronic and electronic components had resulted in a very stable baseline with low temperature coefficient (0.000 87). The sensor response exhibited excellent SNR and reversibility, and had been corrected for both temperature and pressure.

Cerium oxide (ceria) nanoparticles (NPs) have been proved to be an efficient optical fluorescent material through generating visible emission (~530 nm) under violet excitation. This feature allowed ceria NPs to be used as an optical sensor via the fluorescence quenching technique. Shehata *et al* used the gold-ceria NPs for sensing dissolved oxygen (DO) in aqueous media (Shehata *et al* 2020). It was observed that both fluorescence intensity and lifetime were changed due to increased concentration of DO. Added gold was found to enhance the sensitivity of ceria to DO quencher detection. This enhancement was due to optical coupling between the fluorescence emission spectrum of ceria with the surface plasmonic resonance of gold nanoparticles. In addition, gold caused the decrease of ceria nanoparticles' bandgap, which indicates the formation of more oxygen vacancies inside the non-stoichiometric crystalline structure of ceria. The optically-sensitivity-enhanced ceria nanoparticles due to embedded gold nanoparticles can be a promising sensing host for dissolved oxygen in a wide variety of applications including biomedicine and water quality monitoring.

In terms of sensitive film elements, fluorescence quenching dissolved oxygen sensors can improve the sensor performance by selecting different fluorescence materials and film substrate materials. Common fluorescent substances include pyrene, pyrene butyric acid, and fluoranthene, as well as other polycyclic aromatic compounds. Such fluorescence-sensitive substances do not consume dissolved oxygen, have a high response rate (less than 50 ms), and are relatively stable. To improve the detection sensitivity, ruthenium–chromium complexes, as well as platinum phosphor porphyrins such as platinum tetrafluorophenyl porphyrins (PtTFPP) and platinum octaethyl porphyrins (PtOEP), are also widely used as fluorescence indicators in optical oxygen sensors to detect the concentration of dissolved oxygen. The fluorescence intensity of these types of metal compounds corresponds to the dissolved oxygen partial pressure, and the atomic excited state of these metal compounds has a long life and is very stable, making these compounds ideal fluorescence sensitivity indicators. Chen briefly summarized the common matrix materials used in various studies and proposed a type of dissolved oxygen sensor coated with a Pt(II) complex medical polymer film (Chen *et al* 2011). The sensor used a cylindrical core waveguide structure, which showed the characteristics of rapidity and a low SNR when measuring the dissolved oxygen in a large dynamic range. Silva *et al* extracted chlorophyll a from cabbage and combined it with zinc to form a complex, which was used in an optical sensor with high sensitivity, and the R^2 of fluorescence intensity and voltage inhibition reached 0.9809.

The principle of optical oxygen sensors is the oxygen-quenching effect of the fluorescence of the luminescent molecules (luminescent body) fixed in the matrix;

thus, the choice of substrate material is also an important factor impacting the performance of the sensor. At present, many materials are used as matrixes, including silicone rubber, silica gel, sol–gels, and polymers, most of which have a high oxygen permeability, good mechanical and chemical stability, and excellent optical transparency. Peng *et al* proposed a new method for the determination of dissolved oxygen using a porous optical fiber (Peng *et al* 2014). Sensitive films containing ruthenium–bipyridine ($[Ru(DPP)_3]Cl_2$) were modified on the inner pore walls of the porous fiber using sol–gel film-forming technology, and the test probe was then prepared. The measurement sensitivity of the experiments was 3.6 ppm, the sensor response curve measured in the range of 0–20 mg L^{-1} was approximately linear, and the sensor had a quick response time of 200 ms. Noor *et al* utilized sol–gel technology to dip and coat oxygen indicators on the distal end of the optical fiber, and the designed sensor obtained consistent sensitivity results and a reliable drift performance.

2.3.3 pH

pH value is a very important concept and parameter in a solution chemical environment. The measurement of pH value is very necessary in modern industry, agriculture, biomedical engineering, environmental science, and other research and application fields. In recent years, people continue to develop new pH sensors, among which fiber-optic pH sensors have attracted the interest of researchers all over the world due to their various advantages. The measurement principle of the sensor is based on the law that some chemicals show different optical characteristics in different pH environments. The probe part of the sensor is its core, which is usually a film fixed on the end of the optical fiber, or a molecular probe directly fixed on the inner core of the optical fiber. The absorption and emission spectral characteristics of these chemical indicators fixed on the optical fiber are sensitive to the pH value of the external environment, which varies with the change of the pH value of the solution. This change is collected and transmitted by the optical fiber and finally converted into a digital signal, so as to read the pH value of the solution. The fiber-optic pH sensor is not only highly sensitive, but also can be used to measure the pH value under various conditions through the replacement of the probe. It can realize continuous and automatic measurement, and has a broad application prospect and value in the field related to the pH value of solution.

More and more kinds of pH sensitive indicators are used as probes of optical fiber pH sensors, which greatly expand the application space of pH sensors, and make breakthrough progress in the analysis and measurement of pH values in various solution environments. In these studies, naphthalimide, coumarin derivative, Schiff base, ruthenium fluorescence, acridine derivative, quantum dot fluorescence, cellulose pH sensitive and some other fluorescence sensitive substances are used as probes.

A nano-pH sensor based on dual-fluorophore-doped nanoparticles that offer the advantages of adequate sensitivity, accuracy and rapid detection of pH was proposed by Gao *et al.* Additionally, this sensor showed excellent stability, high reproducibility and a fast response time. The development of a pH-sensitive fluorescence nano-sensor by co-doping the indicator dye and reference dye in silica nanoparticles is considered due to their biocompatibility (Gao *et al* 2007).

Novel complex polymers with improved pH sensing properties are constantly emerging, thanks to the hard work of chemists all over the world. As an example, one may refer to the paper by Bojinov *et al*, where the fluorescent properties of a number of polymers were explored in terms of their pH sensitivity. Figure 2.3 shows the changes in the fluorescence emission spectra of 1,2,2,6,6-pentamethylpiperidin-4-ol as a function of pH (Bojinov *et al* 2008).

Goncalves *et al* used complexes [Ru(Phen)$_2$Dcbpy]Cl$_2$ and [Ru(Ph$_2$Phen)$_2$Dcbpy]Cl$_2$ as pH sensitive optics indicator (Goncalves *et al* 2008). It was found that the fluorescence intensity of the compound changed obviously in the range of pH = 3–9, and the fluorescence lifetime was about a few hundred nanoseconds, indicating that these ruthenium(II) complexes are good pH sensitive indicators. The spectral characteristics of pH response in water–alcohol solution (28% ethanol) and sol–gel immobilized in the tip of optical fibers had been characterized in detail. It was found that the fluorescence lifetime of the probe with immobilized complex was improved, which is suitable for accurate measurement in the range of pH = 3–9. The optical fiber pH sensor is simple, convenient, cheap, and practical.

Safavi *et al* developed an optical pH sensor based on immobilization of a mixture of two dyes (dipicrylamine and Victoria blue) on a triacetylcellulose membrane (Safavi and Bagheri 2003). Application of a back-propagation artificial neural network (ANN) model extended the measuring range of the proposed optode to the whole pH range. This sensor exhibited advantages over many existing optical pH sensors including wider dynamic pH range, ease of fabrication, good reversibility, and stability.

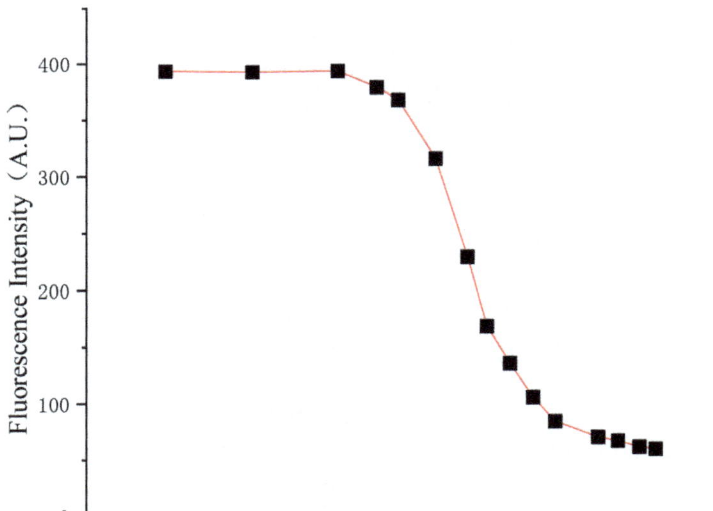

Figure 2.3. Effect of pH on fluorescence intensity (λ_{ex} = 370 nm) of 1,2,2,6,6-pentamethyl-piperidin-4-ol.

2.3.4 Ammonia nitrogen

The sum of ammonium ion and free ammonia in the environment is called ammonia nitrogen. Ammonia nitrogen is the main source of nitrogen essential for the growth of aquatic organisms. Whether its content is too high or too low, it will cause damage to the ecosystem. Most water quality monitoring items include regular on-site sampling, sample preservation, transfer, laboratory analysis, data processing, and other links, which consume a lot of manpower, material resources and time, and the concentration of the analyzed target substance may change after sample preservation and transfer; on the other hand, with a discrete sampling method it is easy to ignore accidental pollution events. Therefore, it is particularly important to develop on-site detection methods of ammonia nitrogen.

The fluorescence method for the reaction of ammonium with OPA has attracted much attention due to its high sensitivity. The interference of amino acids should be considered in the application of ammonium analysis because the reaction can simultaneously determine ammonium and amino acids. In order to improve the selectivity of this method, the reaction was improved by replacing mercaptoethanol with sulfite, so that the sensitivity and selectivity of ammonium were better than that of amino acids (Zhang and Dasgupta 1989). Based on the reaction of NH_3–OPA–sulfite and the application of SFA technology, Kerouel *et al* (1997) proposed a fully automatic direct fluorescence analysis method to measure the concentration of ammonia nitrogen in seawater and estuary. This method eliminates the interference of primary ammonia, and the salinity of 0.2–35 ppt has no effect on it. Li *et al* (1999) studied the fluorescence detection method of ammonia nitrogen using the liquid core waveguide combined with FIA technology. Zhu *et al* (2013) developed a flow batch detection system with solid phase extraction and fluorescence analysis for the determination of trace ammonia nitrogen less than 1 nM in seawater. The fluorescence reaction products were extracted by solid phase extraction column and separated from seawater matrix to realize the enrichment of ammonia nitrogen. Then the compounds were extracted by ethanol and determined by fluorescence method. The limit of detection of this method is 0.7 nM and 1.2 nM, respectively, when it is used in land-based and shipboard laboratory analysis, and there is no interference of salinity and matrix.

The fluorescence reaction of NH_3–OPA–sulfite is very sensitive to pH value, and the optimum pH value is 11. However, under this pH value, hydroxide precipitation is easily formed due to the presence of metal ions in the water sample to be measured. Hu *et al* (2014) found that there was no precipitation when the pH value was adjusted to 11 with EDTA–NaOH buffer, so a modified *o*-phthalaldehyde fluorometric analytical method was established. A new fluorescent reagent 4-methoxyphthalaldehyde (MOPA) was synthesized by Liang *et al* (2015). MOPA can react with ammonium ion rapidly under alkaline and room temperature conditions, so a portable fluorescence detection system was developed. Zhang *et al* (2018) developed a portable laser diode fluorescence detection system using a new fluorescent reagent 4,5-dimethoxy-phthalaldehyde (M_2OPA) as probe. Wang *et al* (2018) improved the circuit design and software algorithm of the fluorescence detector, and introduced a pseudo-random sequence into fluorescence measurement for the first time to realize the extraction of weak fluorescence signal; the detector can also realize self-calibration. Dong *et al* (2017) developed a new ammonia nitrogen detection method based on the volatility of ammonia and the sensitivity of pH change caused by the introduction of silver nanoparticles, using silver nano-clusters as probes and temperature gradient HS-SDME technology.

2.3.5 Nitrite and nitrate

Nitrogen is one of the main factors causing eutrophication in water; it can be divided into organic nitrogen and inorganic nitrogen. Among them, nitrate nitrogen and nitrite nitrogen are the forms of inorganic nitrogen, which are widely found in water. They are mainly caused by the excessive use of agricultural fertilizers and the discharge of wastewater from human living and other production activities. However, the presence of excessive nitrate and nitrite in water has a great impact on aquatic organisms and human health.

When the content of nitrite nitrogen exceeds 0.02 mg L^{-1}, which is directly toxic to aquatic organisms, it affects their growth and development, induces outbreaks of diseases and even causes a large number of deaths. Nitrite in drinking water can irreversibly convert hemoglobin into methemoglobin, which results in a decrease in oxygen carrying capacity in the blood. It can react with secondary amines to produce *n*-nitrosamines, which are likely to lead to gastric cancer. Since nitrate is converted into nitrite in the digestive tract, it also very harmful to the human body. It can also lead to the unnatural reproduction of aquatic plants and algae, leading to 'red tides' and the death of fish. The United States (U.S.) Environmental Protection Agency has limited the nitrite and nitrate content in drinking water to 1 mg L^{-1} and 10 mg L^{-1} and other countries have made similar regulations. The concentration of nitrate and nitrite in water has become one of the indices for evaluating water quality. Therefore, the detection of nitrate and nitrite in water samples at low concentrations has been the focus of research. Usually nitrite (NO_2^-) is detected initially, then nitrate (NO_3^-) is detected mainly by being reduced to nitrite (e.g., Cu/Cd column). In the process of detecting NO_2^- by fluorescence spectrometry, diazotization or nitrosation reaction mainly occurs with non-fluorescent substances (acetaminophen, 2,3-diaminonaphthalene, *n*-methyl-4-hydrazino-7-nitrobenzofurazan, 1,3-naphthalenedisulfonic acid, 5,6-diamino, etc) to produce fluorescent compounds, which can be detected by fluorescence spectrometry.

The principle of detecting NO_2^- based on diazotization reaction is to use aromatic primary amines to react with NO_2^- to form diazo salt, which reacts under acidic conditions, resulting in the change of fluorescence peak. Dong and Dai (1999) used 2,3-diaminonaphthaleneto react with NO_2^-. In the system of DAN–NO_2^-–NaOH, the acidity of the reaction system should be adjusted to 1.64, and surfactant should be added in the whole reaction system to enhance the fluorescence intensity. This method was used to detect NO_2^- in groundwater, and the results showed that this method had high recovery, but the stability is poor and more reagents are needed. The detection of NO_2^- based on nitrosation is the rapid reaction of NO_2^-

with phenols under acidic conditions to obtain colored nitroso derivatives, in which NO_2^- is used as nitrosation reagent. Ohta *et al* (1986) observed fluorescence in alkaline medium by nitrosation of 4-hydroxycoumarin in acidic medium followed by reduction to 3-amino-4-hydroxycoumarin. The results show that the higher the concentration of NO_2^- is, the greater the fluorescence intensity is. The concentration range of NO_2^- in the sample is from 1 pg mL^{-1} to 3 ng mL^{-1} solution, and the relative standard deviation is 0.5%.

There are many kinds of fluorescent probes, such as quantum dots, organic dyes and so on. Quantum dots as fluorescent probes have high photochemical stability, and organic dyes as fluorescent probes have excellent chemical degradation resistance and fluorescence properties. Liu *et al* (2013) prepared near-infrared fluorescent gold nanoclusters (AuNCs) by wet chemical method for the detection of NO_2^-. The fluorescence of AuNCs can be selectively quenched by NO_2^-. With the increase of NO_2^- concentration, the fluorescence intensity decreases linearly in a wide range from 2.0×10^{-8} mol L^{-1} to 5.0×10^{-5} mol L^{-1}. The detection limit was 1.0×10^{-9} mol L^{-1}. Chen *et al* (2011) used hPEI–AgNCs for sensitive detection of nitrite, the mechanism is that under acidic conditions nitrite reacts with H_2O_2 to form peroxynitrite, which can induce the aggregation of hPEI–AgNCs and reduce the fluorescence intensity. The limit of detection of the sensor is 100 nmol L^{-1}, which can be used to detect nitrite in seawater and other practical samples. Gold nanoclusters, as fluorescent probes, have attracted special attention due to their low toxicity, excellent biocompatibility, stability, good solubility and excellent luminescent properties. Du *et al* (2018) synthesized β-CD@AuNPs functional gold nanoparticles with β-cyclodextrin and gold nanoparticles, and then used them together with neutral red as fluorescent probe NR-β-CD@AuNPs to form a sensitive NO_2^- fluorescence sensor. Neutral red (NR), a heteroanthracycline dye, can produce fluorescence under the excitation light. The primary amine group of NR can diazotize with NO_2^- under acidic conditions, resulting in fluorescence quenching, so as to detect the existence of NO_2^-. The results show that this method has high recovery rate and the relative standard deviation is less than 5%. The detection of NO_2^- by fluorescence spectrometry has the characteristics of low dosage, high sensitivity and easy detection, but it is sensitive to environmental factors and has many interference factors such as Fe (III), V (V), and Cr (VI).

Based on the enhanced chemiluminescent reaction signal between diazotized luminol with H_2O_2, a novel chemiluminescent method with high selectivity for the detection of NO_2^- has been established. Combined with flow injection technology, the simultaneous detection of nitrite and nitrate in aqueous solution has been realized (Yaqoob *et al* 2012). Firstly, the content of original nitrite was detected, then the total amount of nitrite reduced by copper plated cadmium column was detected, and the content of the two substances was obtained by difference calculation. Some scholars (Kodamatani *et al* 2009) used anion exchange column to separate nitrite and nitrate in flow injection system. The two anions were transformed into peroxynitrite by low-pressure mercury lamp on-line ultraviolet irradiation. Finally, the simultaneous determination of nitrite and nitrate was realized by mixing reaction with luminol reagent. The detection sensitivity of nitrite is about 10 times higher than that of nitrate. It can be successfully applied to various water samples, such as river water, pond water, rainwater, commercial mineral water, and tap water.

2.3.6 Heavy metals

Heavy metals widely exist in nature, mainly including mercury (Hg), copper (Cu), and lead (PB). Even at a very low concentration, mercury has a high toxicity. It is mainly absorbed into the body through the digestive tract, respiratory tract and skin, causing substantial damage to the nerves and endocrine system. Mercury ion has a high affinity for the sulfhydryl group in the protein, causing many health problems in the brain, kidney and central nervous system, such as goiter and Minamata disease. Copper ion is a significant metal pollutant. According to the EPA, the upper limit of copper ion content in drinking water is 1.3 ppm (~20 μmol L^{-1}). Studies have shown that the cytotoxic effects of copper ions can lead to serious diseases, such as Indian child cirrhosis (ICC), prion disease, amyotrophic lateral sclerosis and Wilson's disease. Lead ion is a kind of neurotoxin harmful to the human body. It enters the body through the food chain, accumulates in the body and causes acute and chronic poisoning, which greatly interferes with the metabolism of the body.

Molecular fluorescence probe technology is a rapid detection technology of heavy metals, which has become a powerful tool for sensing and imaging, and is widely used in environmental and biological samples. Among them, organic dye coumarin, rhodamine, quinoline and their ratio type as representative fluorescent molecular probes have important applications in the rapid detection and field analysis of heavy metals in drinking water and other water quality samples.

The fluorescent probe is generally composed of two basic elements: the reporter and the receiver. The substrate interacts with the probe receptor unit through various forces (such as intermolecular hydrogen bond, electrostatic interaction, covalent bond, etc) to change the fluorescence signal of the signal reporting unit, so as to realize the sensing and detection of the target. The mechanisms of signal transmission include: photoelectron transfer (PET), fluorescence resonance energy transfer (FRET), and intramolecular charge transfer (ICT).

Xu *et al* designed and synthesized a novel coumarin based mercury ion fluorescence probe, namely a thiosemicarbazide derivative, which can detect Hg^{2+} in mammalian cells or *Escherichia coli.* Hg^{2+} induced the desulfurization of thiosemicarbazide derivatives. The green fluorescence was enhanced at 495 nm, and the red shift of the maximum emission wavelength was 10 nm, indicating the occurrence of fluorescence ICT. The new fluorescent probe shows good selectivity and practicability in an ethanol–water solution system, and has a wider application and biological significance (Xu *et al* 2014). Yeh *et al* reported a coumarin fluorescence probe containing phenol hydrazone group. When Cu^{2+} was added, the color of the system changed from light yellow to red, and the intensity of emission peak increased significantly when CN^- was added to the sensor system. When Cu^{2+} was added again, the fluorescence of the probe was quenched again, which confirmed that the binding between the probe and Cu^{2+} is reversible. Even if there is interference of other metal ions, Cu^{2+} is the only metal that causes significant fluorescence quenching. After HeLa cells are treated with Cu^{2+}, the strong fluorescence quenching inside the cells realizes the recognition of Cu^{2+} in the living cells (Yeh *et al* 2014).

Wang *et al* designed a highly selective fluorescence probe based on rhodamine B derivatives. In the presence of Hg^{2+}, the strong affinity of S and N atoms in the probe chelates Hg^{2+}, forming a strong conjugation system. At the same time, the spiro form of the probe was converted to the ring opening amide form, and Hg^{2+} could be recognized in methanol water solution, with fluorescence enhancement and significant color change from colorless to red. The fluorescence intensity of the probe was positively proportional to the concentration of Hg^{2+} (Wang *et al* 2015). Zhang *et al* proposed a series of rhodamine triazine derivatives (probes I–IV) with rhodamine as fluorescence group and cyanuric acyl chloride as molecular platform (Zhang *et al* 2014). The spectral analysis shows that the probes with different structural substitution modes have different sensitivity and selectivity to specific metal ions. Probe I and probe II are active chlorine atoms on the triazine ring replaced by rhodamine B hydrazine, which are fluorescent colorimetric probes for recognizing Cu^{2+}. Under UV light, it can be observed that the orange fluorescent solution turns into pink fluorescent solution, while probe III and probe IV are obtained from rhodamine B ethylenediamine, which can recognize the probes of Al^{3+}, Cr^{3+}, and Fe^{3+}. The introduction of hydroxyl group increases the water solubility of probe II, which has a good recognition of Cu^{2+} in neutral solution. Zhang *et al* summarize the previous work, and propose building a molecular logic gate for the visual transformation of probe II to Cu^{2+} and S^{2-}. The color of the solution changed from pink to colorless, which indicated that S^{2-} substituted for probe II and coordinated with Cu^{2+}, forming a more stable CuS, which confirmed the reversibility of probe II.

Basa *et al* discussed a rare chemosensor and chemodosimeter for multi-ion reaction and single molecular system. Selective reduction of an anthracenone–quinoline imine derivative, using 1.0 equiv of $NaBH_4$ in 95% ethanol afforded the corresponding anthracen-9-ol derivative. Zn^{2+} coordination induced a 1,5-prototropic shift resulting in anthracene fluorophore formation via an imine–enamine tautomerization pathway. Cu^{2+} induced the colorimetric change of the sensor from light yellow to orange red, which resulted in irreversible hydrolysis of imine and complete quenching of fluorescence (Basa and Sykes 2012).

Zhou *et al* designed a simple fluorescence probe containing rhodamine and aminoquinoline groups (Zhou *et al* 2013). According to the job plot experiment, the probe and Hg^{2+} are connected by 1:1 stoichiometry. When Hg^{2+} was added into the probe solution gradually, the emission peak of the probe (5×10^{-5} mol L^{-1}) at 497 nm decreased gradually, accompanied by a new peak formation at 581 nm, and the fluorescence intensity at 581 nm gradually increased with the increase of Hg^{2+} concentration, and then reached equilibrium, accompanied by the change from colorless solution to red solution, emitting bright green fluorescence. At 420 nm, FRET may occur in the probe when quinolone–zinc complex is excited. In addition, the ratio change of fluorescence spectrum separated the emission point at 547 nm.

Gong *et al* designed and synthesized a ratio probe of coumarin–rhodamine TBET system (Gong *et al* 2012). Through the use of Hg^{2+} desulfurization reaction as the recognition mechanism, it shows strong selectivity and high sensitivity to Hg^{2+}. The blank probe ($[Hg^{2+}] = 0$ mol L^{-1}) ha^{-1} s^{-1} an obvious absorption peak at 420 nm, accompanied by a yellowish color emitted by the donor (coumarin), and has no receptor absorption peak characteristic (rhodamine B, nearly 560 nm), which proves the existence of the structure in the snail. However, when Hg^{2+} was added into the buffer solution, the absorption band of the donor showed a red shift. This result may be due to the fact that Hg^{2+} triggers the transformation of thiosemicarbazide to 1,3,4-oxadiazole, which increases the degree of conjugation, indicating a strong electron absorption effect on coumarin. In addition to the red shift of coumarin absorption peak, a new characteristic absorption peak corresponding to the open-loop structure of rhodamine B appears at 567 nm, which has a high molar extinction coefficient. With the increase of Hg^{2+} concentration, the absorption intensity increases. At the same time, it is easy for the naked eye to observe the significant change of the sensor color from yellowish to red, and the change of fluorescence from cyan to pink (excited at 365 nm).

2.4 Conclusion and future perspectives

Fluorescence analysis has been widely used in the monitoring of chlorophyll, CDOM and dissolved oxygen in water, and corresponding commercial sensors have been formed. However, its application in pH, ammonia, nitrate, nitrite, and heavy metals in water is still at the level of laboratory research. In recent years, the research of fluorescent probe used to detect pH, ammonia nitrogen, nitrate nitrogen, nitrite nitrogen, and heavy metals in water has developed rapidly. The design and synthesis of new materials make the above-mentioned detection objects continuously expand in the fields of environmental monitoring, food detection, and biological toxicology. The matrix of real samples is complex, and the fluorescent probe is easy to be interfered by many kinds of detection environment, so its detection accuracy and utilization rate are low. Therefore, it is an important research content to break through the traditional design concept, build a multi-functional and high-performance detection system, and develop a new detection mechanism. At the same time, keeping the excellent detection performance of sensor, developing reusable, environment-friendly and real-time monitoring sensor is also a research hotspot at present and in the future. In a word, many kinds of technologies are cross-fused to learn from each other. The design and synthesis of a new type of water quality sensor based on fluorescence method will come in the near future.

Includes extracts reproduced from the authors' previously published papers (Wei *et al* 2019, Li *et al* 2020). CC BY 4.0.

References

Basa P N and Sykes A G 2012 Differential sensing of Zn(II) and Cu(II) via two independent mechanisms *J. Org. Chem.* **77** 8428–34

Blockstein L and Yadid-Pecht O 2014 Lensless miniature portable fluorometer for measurement of chlorophyll and CDOM in water using fluorescence contact imaging *IEEE Photon. J.* **6** 1–16

Bojinov V B, Simeonov D B and Georgiev N I 2008 A novel blue fluorescent 4-(1,2,2,6,6-pentamethylpiperidin-4-yloxy)-1,8-naphthalimide pH chemosensor based on photo induced electron transfer *Dyes Pigm.* **76** 41–6

Chen R, Farmery A D, Obeid A and Hahn C E 2011 A cylindrical-core fiber-optic oxygen sensor based on fluorescence quenching of a platinum complex immobilized in a polymer matrix *IEEE Sens. J.* **12** 71–5

Dong J and Dai J P 1999 Research and application of fluorimetric determination of nitrite *Shandong Environ.* **4** 3–4 (in Chinese)

Dong J X, Gao Z F and Zang Y A 2017 Selective and sensitive optical sensor for dissolved ammonia detection via agglomeration of fluorescent Ag nanoclusters and temperature gradient head space single drop microextraction *Biosens. Bioelectron.* **91** 155–61

Du X, Zhang X, Jiang C, Zhang W and Yang L 2018 The trace detection of nitrite ions using neutral red functionalized SH-β-cyclodextrin @Au nanoparticles *Sensors* **18** 681–92

Gao F, Tang L, Dai L and Wang L 2007 A fluorescence ratiometric nano-pH sensor based on dual-fluorophore-doped silica nanoparticles *Spectrochim. Acta A: Mol. Biomol. Spectrosc.* **67** 517–21

Goncalves H M R, Maule C D, Jorge P A S and Silva J C G E D 2008 Fiber optic lifetime pH sensing based on ruthenium(II) complexes with dicarboxybipyridine *Anal. Chim. Acta* **626** 62–70

Gong Y J, Zhang X B and Zhang C C *et al* 2012 Through bond energy transfer: a convenient and universal strategy toward efficient ratiometric fluorescent probe for bioimaging applications *Anal. Chem.* **84** 10777–84

Hu H Z, Liang Y, Li S, Guo Q and Wu C C 2014 A modified *o*-phthalaldehyde fluorometric analytical method for ultratrace ammonium in natural waters using EDTA–NaOH as buffer *J. Anal. Methods Chem.* **2014** 1–7

Hu Y, Pan Y and Sun L *et al* 2013 An in-situ highly sensitive and reliable chlorophyll sensor based on pseudo-random sequence modulation *IEEE Trans. Instrum. Meas.* **62** 2314–22

Kodamatani H, Yamazaki S, Saito K, Tomiyasu T and Komatsu Y 2009 Selective determination method for measurement of nitrite and nitrate in water samples using high-performance liquid chromatography with post-column photochemical reaction and chemiluminescence detection *J. Chromatogr.* A **1216** 3163–67

Kerouel R and Aminot A 1997 Fluorometric determination of ammonia in sea and estuarine waters by direct segmented flow analysis *Marine Chem.* **57** 265–75

Lamb J J, Eaton-Rye J J and Hohmann-Marriott M F 2012 An LED-based fluorometer for chlorophyll quantification in the laboratory and in the field *Photosynth. Res.* **114** 59–68

Li D, Wang T, Li Z, Xu X, Wang C and Duan Y 2020 Application of graphene-based materials for detection of nitrate in water—a review *Sensors* **20** 54

Li J, Dasgupta P K and Zhang G F 1999 Transversely illuminated liquid core waveguide based fluorescence detection: fluorometric flow injection determination of aqueous ammonium/ammonia *Talanta* **50** 617–23

Liang Y, Pan Y M, Guo Q, Hu H Z, Wu C C and Zhang Q 2015 A novel analytical method for trace ammonium in freshwater and seawater using 4-methoxyphthalaldehyde as fluorescent reagent *J. Anal. Methods Chem.* **2015** 1–7

Liu H, Yang G and Abdelhalim E S 2013 Highly selective and ultrasensitive detection of nitrite based on fluorescent gold nanoclusters *Talanta* **104** 135–39

McDonagh C, Kolle C, Mcevoy A K, Dowling D L, Cafolla A A and Cullen S J *et al* 2001 Phase fluorometric dissolved oxygen sensor *Sens. Actuators* B **74** 124–30

Ohta T, Arai Y and Takitani S 1986 Fluorometric determination of nitrite with 4-hydroxycoumarin *Anal. Chem.* **58** 3132–35

Peng L, Yang X, Zhang J, Dai L, Ma Z and Peng S 2014 Fluorescence detection method of dissolved oxygen based on fiber *Chin. J. Lumin.* **35** 382–86

Rao G, Lam H, Kostov Y, Tolosa L and Ge X 2011 *Chlorophyll and turbidity sensor system* US20110273705

Safavi A and Bagheri M 2003 Novel optical pH sensor for high and low pH values *Sens. Actuators* B **90** 143–50

Sharikova A V and Killinger D K 2010 Laser and UV-LED-induced fluorescence detection of dissolved organic compounds in water *Proc. SPIE* **7666** 76661L

Shehata N, Kandas I and Samir E 2020 In-situ gold-ceria nanoparticles: superior optical fluorescence quenching sensor for dissolved oxygen *Nanomaterials* **10** 314

Wang C, Li Z, Pan Z L and Li D L 2018 Development and characterization of a highly sensitive fluorometric transducer for ultra-low aqueous ammonia nitrogen measurements in aquaculture *Comput. Electron. Agric.* **150** 364–73

Wang F H, Cheng C W and Duan L C *et al* 2015 Highly selective fluorescent sensor for Hg^{2+} ion based on a novel rhodamine B derivative *Sens. Actuators B: Chem.* **206** 679–83

Watras C J, Hanson P C, Stacy T L, Morrison K M, Mather J and Hu Y H *et al* 2011 A temperature compensation method for CDOM fluorescence sensors in freshwater *Limnol. Oceanogr. Methods* **9** 296–301

Wei Y, Jiao Y, An D, Li D, Li W and Wei Q 2019 Review of dissolved oxygen detection technology: from laboratory analysis to online intelligent detection *Sensors* **19** 3995

Xu Y, Jiang Z and Xiao Y *et al* 2014 A new fluorescent turn-on chemodosimeter for mercury ions in solution and its application in cells and organisms *Anal. Chim. Acta* **807** 126–34

Yaqoob M, Biot B F, Nabi A and Worsfold P J 2012 Determination of nitrate and nitrite in freshwaters using flow-injection with luminol chemiluminescence detection *Luminescence* **27** 419–25

Yeh J T, Chen W C and Liu S R *et al* 2014 A coumarin-based sensitive and selective fluorescent sensor for copper(II)ions *New J. Chem.* **38** 4434–39

Yin G, Zhao N and Shi C 2018 Phytoplankton photosynthetic rate measurement using tunable pulsed light induced fluorescence kinetics *Opt. Express* **26** A293–300

Zhang G F and Dasgupta P K 1989 Fluorometric measurement of aqueous ammonium ion in a flow injection system *Anal. Chem.* **61** 408–12

Zhang M, Zhang T, Liang Y and Pan Y M 2018 Toward sensitive determination of ammonium in field: a novel fluorescent probe, 4,5-dimethoxyphthalaldehyde along with a hand-held portable laser diode fluorometer *Sens. Actuators* B **276** 356–61

Zhang Y, Zeng X and Mu L *et al* 2014 Rhodamine-triazine based chemosensors for Cu^{2+} in aqueous media and living cells *Sens. Actuators B: Chem.* **204** 24–30

Zhou X, Yan W and Zhao T *et al* 2013 Rhodamine based derivative and its zinc complex: synthesis and recognition behavior toward Hg(II) *Tetrahedron* **69** 9535–39

Zhu Y, Yuan D, Huang Y, Ma J and Feng S 2013 A sensitive flow-batch system for on board determination of ultra-trace ammonium in seawater: method development and shipboard application *Anal. Chim. Acta* **794** 47–54

Chapter 3

Paper-based optical sensors for water analysis and monitoring

Bambang Kuswandi

Classical analytical methods used for water quality monitoring are often laborious, time-consuming, and complicated, and need trained personnel to perform the analysis. Hence it is important to develop analytical methods that are simple, rapid, and low-cost. For this purpose, the development of a paper-based sensor could act as an alternative approach that can be employed for water analysis and monitoring. Paper-based sensors are also referred to as microfluidic paper-based analytical devices (μPADs) or lab-on papers, where they provide simple, rapid, low-cost, and portable devices for water analysis and monitoring. In this chapter, the development of paper-based optical sensors is discussed for application in water analysis and monitoring, including highlighting the features of these devices (paper types and fabrication, optical sensor types, and modes) and their applications for detection of different pollutants in water samples. It can be stated that these devices are showing potential to simplify and improve the analytical method performance for water pollutant detection, particularly if the optical signal of the sensor is integrated with machine learning and artificially intelligent, for its application for water quality analysis and monitoring.

3.1 Introduction

Globally, water pollution has become a great concern, since it is a serious environmental problem worldwide. Since industrialization, population, and urbanization are increasing in most countries around the world, undoubtedly degradation of water quality will be affecting human health, and ecosystem quality, reducing food production, and economic growth. Currently, water quality degradation occurs continuously caused by major pollutants that have been released in large amounts into water systems, e.g., rivers, lakes, coasts, etc. This pollution issue is increased further by the emerging continuous discharge of pollutants, such as pharmaceuticals,

doi:10.1088/978-0-7503-3047-3ch3

cosmetics, organic pollutants, and others that give a more negative impact on human health and the environmental system (Schwarzenbach *et al* 2010, Evans *et al* 2019).

To reduce the negative impact, and cost related to water pollution, a continuous water quality monitoring system is critical to protect environmental water systems and ensure safe water supplies for societies. It is therefore crucial to monitor a variety of water pollutants where information can be used for preventing and controlling further water pollution. Based on this point of view, the challenge is mainly to fulfill the need for analytical tools with high accuracy and sensitivity that can be used to analyze a wide range of water pollutants using a simple and rapid method (Jéquier and Constant 2010, Koop and Van Leeuwen 2017).

In this context, paper-based sensors provide versatile approaches as a sensor platform, as they are flexible, simple, and have a high surface area as well as the internal ability to separate and filter target analytes via osmotic force. Moreover, they have good permeability to liquid and gas, and low cost (Kung *et al* 2019, Liana *et al* 2012, Singh *et al* 2018). In this context, paper-based sensors are also referred to as microfluidic paper-based analytical devices (μPADs) (Almeida *et al* 2018), paper-based analytical devices (PADs) (López-Marzo and Merkoçi 2016) or lab-on papers (Chitnis *et al* 2011), To develop paper-based sensors, fabrication with low-temperature techniques is needed, which in turn needs new fabrication processes and functional materials, such as printing, synthesis, and immobilization techniques. One of the major benefits of employing paper as a sensing platform is that it allows for miniaturization that only needs low reagent and sample volume, low cost for mass production and simple sensor fabrication (Kung *et al* 2019, Singh *et al* 2018).

Paper as a sensing platform for optical chemical sensor and biosensor development allows detection systems using different sensor designs and fabrication techniques that could be applied (Singh *et al* 2018). This chapter discussed the different sensor designs, distinct fabrication techniques, including different modifications approached, in achieving a reliable, sensitive, and reproducible signal. Moreover, various materials used in the modification of the sensor surface to enhance the signal sensitivity for the target analyte are discussed as well. As a sensing platform, paper is an ideal platform to develop optical chemical and biosensors that are user-friendly, cost-effective, and sensitive, which have a high potential for customization depending on the application required. Commonly paper-based sensors as analytical devices are fabricated to be simple, portable, and disposable, and therefore well-suited for field applications. However, the application of paper-based optical sensors for water analysis is increasing currently. Hence, the objectives of this chapter are to provide information on current developments in paper-based optical sensors and their applications to detect pollutants in water, particularly for the works that have been developed recently, including possibilities to used machine learning and artificial intelligence, for their application for water quality analysis and monitoring.

3.2 Paper-based sensors

3.2.1 Paper types

Paper mainly consists of cellulose fibers with a complex structure that allows them to be used as a potential platform for sensor development as analytical devices. As a sensing platform, paper has many advantages, e.g., versatility, biocompatibility, and low cost, including simple and disposable analytical tools with low reagent and sample volume (microliters) (Busa *et al* 2016, Hosseini *et al* 2017, Liana *et al* 2012). Since it is mainly cellulose fibers, it is a polymer with hydrophilic properties that is permeable to aqueous liquids or water (Almeida *et al* 2018). The various types of paper that have been used for paper-based sensor development is given in table 3.1.

Filter papers have been commonly used as solid support material for sensors application (Ahmed *et al* 2016, Almeida *et al* 2018, Akyazi *et al* 2018, Liana *et al* 2012, Sher *et al* 2017). They are commercially available in a range of various filter paper grades, that mostly consist of highly stable cellulose, i.e., alpha-cellulose. They can be classified based on different properties, e.g., thickness, pore size, flow rate, and particle retention. Filter paper grade 1 is classified as a medium retention and flow filter paper, and it has been functionalized to obtain lateral flow analysis or paper-based immunosensors (Posthuma-Trumpie *et al* 2009, Xie *et al* 2015). It also has been developed for heavy metal detection by adsorbing metallothioneins (Irvine *et al* 2017). Similarly, it has been used for the incorporation of an *in vitro* transcription or translation system (Duyen *et al* 2017).

Cellulose chromatography paper is a cellulose paper with a high-grade in paper-based sensors. It can be classified by its thickness and flow rate. By taking advantage of its separative properties, a paper-based sensor using chromatographic paper has been used to increase concentration of gold nanoparticles (AuNPs) on the paper surface, where reduced silver particles at low concentrations were detected (Vijitvarasan *et al* 2015). It has also been employed in the development of a microfluidic device based on capillary flow action, where the lower flow rate (115 mm/30 min, grade 2) was chosen for immunoagglutination assay (McCracken *et al* 2017). Cellulose derivatives, e.g., nitrocellulose membranes, have also been employed as a platform for paper-based sensors, particularly in lateral flow assay (LFA). The membranes have hydrophobic

Table 3.1. Paper types used in paper-based sensors.

Paper type	Properties	Example
Cellulose filter paper	Hydrophilic polymer, various pore sizes and thickness available, permeable to aqueous liquids	Whatman® filter paper (grades 1 and 2)
Cellulose chromatography paper	Hydrophilic polymer, allows nanostructures on its surface and nanoparticles separation, permeable to aqueous liquids	Whatman® chromatographic paper (grades 1 and 2)
Nitrocellulose membrane	Hydrophobic polymer, allows for biological elements immobilization by absorption or electrostatic interactions	Millipore Hi-Flow Plus HF240
Copy paper	Hydrophilic polymer, various pore sizes and thickness available, 3D printed possible, providing pattern or microchannels, permeable to aqueous liquids	Multiuse recycled copy paper, printer paper

properties and show themselves to be suitable for enzymes and proteins immobilization via electrostatic interactions (Liana *et al* 2012). For instance, the nitrocellulose membrane has been used for the development of immunosensor for the detection of Cd(II) (López-Marzo and Merkoçi 2016). Furthermore, it also used for the immobilization of antibodies, to create many zones, such as a zone for the probe conjugate, a zone for positive control using BSA, and a zone for control goat antimouse immunoglobulin that is used to detect U(VI) (Quesada-González *et al* 2018). Moreover, chromatographic paper and copy paper were also employed for printing wax wells as a pattern in paper-based sensors (Vijitvarasan *et al* 2015), or to create microplates with low cost (Adkins *et al* 2017).

3.2.2 Printing and fabrication

Different types of printing and fabrication techniques could be employed to create paper-based sensors with variable functions, e.g., barriers to making pattern/microfluidics for sample flow, or create sensing zones, and immobilized sensing zones with biomolecules or reagents. Wax printing technique is a fabrication technique employed to make hydrophobic patterns or barriers that allow sample flow, fluid reservoirs, or sensing zones or microzones (Carrilho *et al* 2009) as given in figure 3.1. Using wax printing, various approaches have been applied (Altundemir *et al* 2017, Carrilho *et al* 2009, Dungchai *et al* 2011, Liana *et al* 2012, Nilghaz *et al* 2019), such as graphic design software (McCracken *et al* 2017, Scala-Benuzzi *et al* 2018a) or stencils (Jokerst *et al* 2012, Adkins *et al* 2017) to create microfluidic channels and areas. The integration of the wax onto the microfluidic channels is patterned by printing the wax onto the filter paper surface following with baking or heating for wax penetration in the paper. However, this technique is only suitable for low-resolution printing, since the printed patterns are affected by heat as the wax melts easily at high temperature.

The screen printing technique is another printing technique that is employed to produce paper-based sensors. This technique is often used not only to create a sensing zone or microzone (Hidayat *et al* 2019) but also to fabricate a paper-based

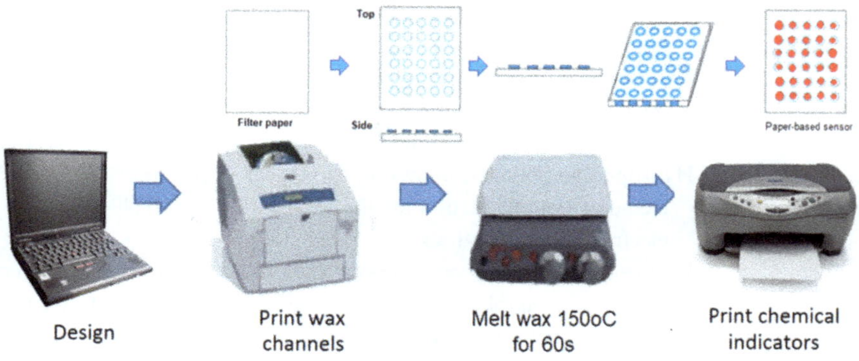

Figure 3.1. Fabrication scheme for wax printing using filter paper to fabricate colorimetric paper-based sensors.

electrode by using a carbon-based conductive ink to print three electrodes onto the surface of the hydrophobic paper (Rengaraj *et al* 2018). Similar fabrication techniques are a fabrication of screen-printed electrodes (Irvine *et al* 2017), accompanied by other simple procedures, such as cutting, etc, (Duyen *et al* 2017). This fabrication procedure often found in lateral flow immunosensors production, and has been developed by assembling different layers that consist of the sample and absorption pad, the nitrocellulose membrane, including the conjugation pad or strip for a colorimetric signal (López-Marzo *et al* 2013, López-Marzo and Merkoçi 2016, Quesada-González *et al* 2018). This printing technique is also not suitable for high resolution printing to create the line patterns or sensing zone (table 3.2).

Photolithography is a printing technique that can be employed to fabricate a high-resolution pattern, and is suitable for paper-based sensor production at a large scale. For example, paper-based sensors that can detect multiple biomarkers were fabricated using this printing technique (Dungchai *et al* 2009). The fabrication processes involve pouring the photoresist onto the paper surface, and distributing it evenly, after which the paper is heated. Then, the paper is masked with a transparent patterned film that is created using a laser printer and irradiated with UV. After heating or baking, using acetone the unpolymerized photoresist is removed. Following the drying process, to make hydrophilic areas, then the paper is exposed to air plasma. This fabrication technique needs expensive equipment and complex and long procedures. Photolithography is also employed without air plasma to create hydrophilic channels for patterned sensors (Klasner *et al* 2010). This technique has been used to fabricate a paper-based sensor for multiple metal complexes detection (Ouyang *et al* 2014). However, paper-based sensors fabricated using this technique can be easily damaged when they are folded or bent.

Table 3.2. The main printing and fabrication techniques used for the paper-based sensors developments.

Techniques	Advantages	Disadvantages
Wax Printing	Simple fabrication, rapid process, low-cost	Low-resolution printing, unstable at high temperature or heating
Screen printing	Low-cost, rapid and simple fabrication, low-cost	Low-resolution printing
Photolithography	High-resolution printing, allows for large-scale fabrication	Need sophisticated equipment for patterning, unstable against folding or bending, expensive
PDMS patterning	Flexible, low-cost	Need sophisticated equipment technique, low resolution
Inkjet Printing	High resolution, efficient, rapid process, low cross-contamination	Expensive ink printer for reagent or bioreagent
Laser Cutting	Simple, rapid process, inexpensive	Need specialized equipment
Hot Embossing	Efficient, rapid process	Need specialized equipment

Since polydimethylsiloxane (PDMS) has hydrophobic properties, printing a PDMS solution on paper offers a solution to fabricate flexible paper-based sensors. A low-cost method using PDMS patterning on paper has been developed to fabricate flexible sensors (Bruzewicz *et al* 2008). This fabrication technique allows producing a paper-based sensor that can be folded or bent without damaging the partners or channels. Here, the PDMS solution with hexanes (3:1 ratio (v/v)), was loaded into a plotter and printed onto the paper to fabricate microfluidics patterns. It has many advantages, such as lower cost than SU-8 photoresist for photolithography, no need for crosslink using UV light, rapid fabrication, and no need for post-printing. However, using this technique, the plotter should be modified using customized cartridges to load the PDMS solution that produces a low-resolution pattern during printing. For example, paper-based biosensors have been developed by patterning PDMS substrates using stencils to capture antibodies (Gervais and Delamarche 2009). A PDMS-based paper has been fabricated by integrating two layers, i.e., channel and patterned layers (Shangguan *et al* 2017). Here, the precursor and the curing agent in a 10:1 (w/w) ratio were applied to prepare PDMS solution, then poured over a stencil to create patterns. The channel and patterned layers were integrated to create microfluidic channels, and cured via heating.

Inkjet printing is a printing technique that allows the fabrication process of paper-based sensors just by the use of an inkjet printer without any additional pieces of equipment. This printing technique is commonly used because of its non-contact process that reduces cross-contamination between reagents and samples. Generally, commercially inkjet printers can be employed with or without modification to print a sensitive sensing zone. For example, a modified inkjet printer and ink cartridges were used for paper-based sensor fabrication by Khan *et al* (2010). Similarly, inkjet printing has been utilized to create electrodes on the paper surface (Gaspar *et al* 2017). Use of inkjet printing allows for the rapid fabrication process of sensors, however, either a modified inkjet printer or an expensive bio-ink printer is needed.

Laser cutting is an inexpensive and simple technique to create patterns or channels on paper substrate. For example, a CO_2 laser cutter has been utilized to make hydrophilic patterns on a paper substrate that has been modified via a hydrophobic coating (Chitnis *et al* 2011). Laser cutting allows converting hydrophobic areas inside and on the surface of paper into hydrophilic areas. The process creates the hydroxyl and carbonyl groups at the paper surface. Similarly, a commercial CO_2 laser cutting has been used to cut through backed and unbacked nitrocellulose membrane, Mylar-based substrates, and glass fibers (Spicar-Mihalic *et al* 2013). It was also utilized to produce small-scale channels in paper-based sensors, for a small sample flow (Mahmud *et al* 2018). Generally, a laser was utilized to create barriers in the paper by stripping the hydrophilic material. Although this technique is simple, it needs specialized equipment, e.g., laser cutter or engraver, double-sided tape, and 2D graphics software, for designing and assembly of paper-based sensors.

Hot embossing is another fabrication technique in the production of the paper-based sensor that creates hollow hydrophilic channels that allow for capillary flow spontaneously. This technique is efficient since its roll-to-roll operation allows for high throughput as low cycle times. A paper-based sensor was developed by embossing PowerCoat® HD paper coated with a rubber layer for hydrophobicity, but hydrophilicity using poly(vinyl alcohol) (Gosselin *et al* 2017). Furthermore, two methods have been introduced in paper-based sensor production via slight or complete parafilm infusing to the paper that resulted in laminated paper (Kim *et al* 2018). Similarly, another method was developed by utilizing parafilm for paper-based sensors (Yu and Shi 2015). Although hot embossing utilization can be efficient, it needs specialized equipment. Another simple fabrication technique has also been developed pre-marking the spots with a pencil, following a cell suspension via physical deposition and drying of the papers (Lisa *et al* 2009).

3.3 Optical sensor types

Different types of optical sensors, i.e., chemical sensors and biosensors, have been successfully developed in the fabrication of paper-based sensors for water analysis and monitoring. This type of paper-based sensor is used as a chemical sensor due to its wicking properties and relatively uniform thickness, including high reagents retention and adsorption that make it suitable for chemical sensors or biosensors (Santhiago *et al* 2014). Due to various physical and/or chemical properties that might be needed for paper-based optical sensors, various types of paper have been studied depending on the target analyte and application. For example, in paper-based biosensor developments, nitrocellulose membranes are widely employed since their structure is suitable for the modification of chemical functional groups that allow for covalent immobilization of biomolecules on the membrane.

3.3.1 Chemical sensors

Chemical sensors are defined as analytical devices or tools that integrate chemical reagent for target analyte detection and a transducer (Kuswandi *et al* 1999). The detection of the target analyte and the transducer converts the chemical reaction into a measurable optical signal (Kuswandi *et al* 2007a). In this context, many reagents, including indicators or reagent dyes and fluorescence reagents, that are commonly used in the optical methods, such as colorimetric, spectrophotometric, and fluorimetric method can be also used in this case (Kuswandi *et al* 2007b). The widely used optical transducers use colorimetric, chemiluminescence, and fluorescence detections. Currently, paper-based sensors have been fabricated using different printing and patterning techniques depending on sensor purposes. Besides the colorimetric detection, other detection techniques, e.g., chemiluminescence, electrochemiluminescence, and fluorescence techniques have been utilized for the detection of target analytes. These approaches have been applied in paper-based optical chemical sensors; due to their advantages, such as simplicity, being user-friendly and rapid detection (Ahmed *et al* 2016), these optical chemical sensors also have limitation related with selectivity and sensitivity toward the target analyte (Liana *et al* 2012). One example of a colorimetric paper-based sensor is a pH indicator paper as given in figure 3.2.

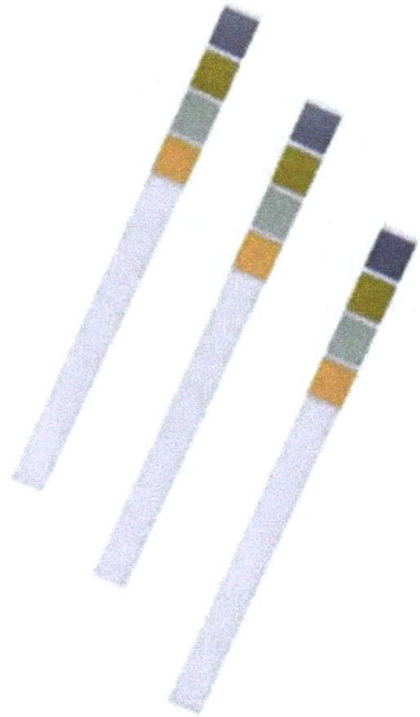

Figure 3.2. Common pH indicator paper based on colorimetric method.

3.3.2 Biosensors

Biosensors can be described as analytical devices or tools that associate or integrate a bioreagent and a transducer. The bioreagent is a bio-element that recognizes the target analyte via the biochemical interaction, where the transducer converts the biochemical reaction into a detectable signal (Kuswandi *et al* 2005). In many biosensors, the transducers used are electrochemical, optical, piezoelectric, and thermal methods. This chapter only focused on the optical transducer type that is widely employed in paper-based biosensors for water analysis and monitoring, e.g., as colorimetric, fluorescence, and chemiluminescence techniques. One example of a colorimetric paper-based biosensor that is widely used in immunochromatography paper or often called laminar flow assay is given in figure 3.3.

In the case of paper used in optical biosensors, the paper should be modified to become bioactive paper. The bioactive paper, usually the paper or cellulose substrate is modified or functionalized with biomolecules. Sometimes the modification of paper, only involves adsorption of cationic interaction of biomolecules onto the wet cellulose matrix or paper. However, this biomolecules adsorption on cellulose, in terms of the rate and extent of the process, is sometimes less effective when compared to the modified hydrophilic surface (Brash and Ten Hove 1993). Another approach also involves chemical modification of cellulose matrices, such as by the paper surface activation via amide or aldehyde, and then covalently binding with biomolecules, so that the cellulose matrix allows absorbing any type of biomolecule. Moreover, this paper modification to create a paper-based biosensor is fabricated via inkjet printing of biocompatible sol–gel layers containing enzymes that can be immobilized onto a paper matrix (Hossain *et al* 2009). The glycical

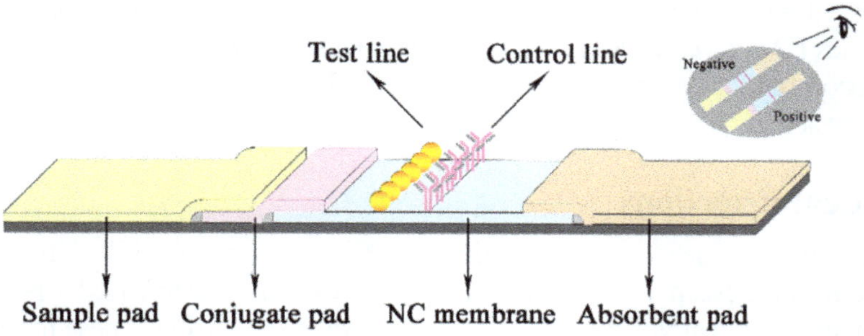

Figure 3.3. The basic design of a colorimetric paper-based biosensor used in immunochromatography, like in pregnancy tests. It consists of a sample pad that can absorb and transport analyte, a conjugated pad that consists of antibodies, where analyte adheres to them, along with a detectable indicator that is highly antigen-specific binds to produce a visible solid line that indicates a positive result. Secondary control lines are used to detect conjugated antibodies and make sure that the test is working properly. Adapted from Pan *et al* 2020. Copyright 2020 by the authors, published by MDPI. CC BY 4.0.

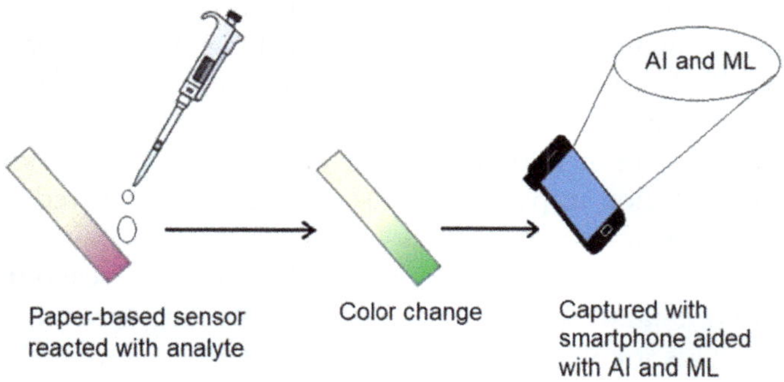

Figure 3.4. Colorimetric detection of paper-based detection using a smartphone camera aided with AI and ML for smart functions, such as autocalibration and accuracy in the measurement of a target analyte.

methacrylate is grafted via copolymerization onto cellulose paper that allows for the biomolecule's immobilization and has been used for gold nanoparticle-based tests and ELISA (Tyagi *et al* 2009).

Even using paper as a matrix, the biochemical interaction that is used as a biorecognition process has a selective and strong affinity to the target analyte. There are different types of biomolecules used, i.e., natural biomolecules (enzymes, antibodies, nucleic acids, and other proteins), synthetic biomolecules (aptamers, and molecularly imprinted polymers), whole cells and microbes (figure 3.4) (Hashemi Goradel *et al* 2018). Various types of biomolecules have been used for the development of the paper-based biosensor for the detection of various target analytes in water samples. For example, the use of antibodies (polyclonal rabbit anti-EE2) in paper-based biosensor has been fabricated for the estrogen ethinylestradiol detection in river water samples (Scala-Benuzzi *et al* 2018b). Moreover, antibody-conjugated particle suspensions have been developed for the detection of the two target pathogens (Zika virus and *Escherichia coli* K12) (Klug *et al* 2018). In other work, enzyme-based paper biosensor has been developed as an on/off switch for the detection of antibiotics (Duyen *et al* 2017). Here, the β-galactosidase enzyme reconstituted from purified recombinant components of *E. coli* is used as a biorecognition element in this biosensor system.

3.4 Optical detections

Different optical modes, such as colorimetric, fluorescence, chemiluminescence, and electrochemiluminescence methods have been used in paper-based optical sensor developments to detect various target analytes (Sher *et al* 2017). These methods have been widely employed as they allow sensitive and selective detection toward the target analyte.

3.4.1 Colorimetric method

Colorimetric methods can be described as methods employed to detect the target analyte presence visually or its concentration level by measuring the color change or color formation by using (a) direct imaging with a camera, smartphones, or scanners combined with an image software program such as ImageJ for quantification (Kuswandi *et al* 2014), or (b) conventional spectrophotometers via sample absorbance measurement at specific wavelengths (Christodouleas *et al* 2015). This method is widely employed since it allows accurate optical detection at low cost. Color change or color formation in paper-based sensors can be triggered using enzymes, AuNPs, or other color tags.

An enzyme based biosensor is based on the catalytic reaction between an immobilized enzyme and specific substrate via an enzyme–substrate complex formation that causes a color change (Yetisen *et al* 2013, 2014). The paper-based biosensor was fabricated via photolithography for proteins and glucose detection in artificial urine (Martinez *et al* 2008). Here, the glucose oxidase enzyme was a catalyst to turn glucose via oxidation to become gluconic acid along with H_2O_2 formation, and is the basis of the biosensing mechanism. Then, using horseradish peroxidase (HRP), H_2O_2 was reduced by iodide oxidation to become iodine, which produced a brown color. For simultaneous proteins detection, tetra-bromophenol blue (TBP) was employed to react with proteins electrostatically to produce color change from yellow to blue. Moreover, competitive enzymatic assay based on nitrocefin as a paper-based biosensor (Boehle *et al* 2018) has been fabricated for β-lactams antibiotics detection, where color turns from yellow to red if antibiotics are present.

Commonly, AuNPs are also used as the colorimetric label for secondary antibodies as detection lines in a paper-based immunosensor often called immunochromatography assays or LFA (Choi *et al* 2010, Chun 2009). In this scheme, a line color formation is produced, due to aggregation of AuNPs with their specific target analyte (e.g., antigen or antibody). The qualitative detection can be performed visually using the naked eye, while quantitative detection can be performed using imaging software so that the color values (e.g., RGB) can be correlated with target analyte concentrations. For example, a paper-based biosensor using gold nanoparticle-tagged antibodies has been developed for tuberculosis detection (Tsai *et al* 2017). AuNPs are used in paper-based colorimetric sensors due to having various advantages, e.g., high surface area and biocompatibility for various affinity interactions. However, upon washing, AuNPs could leak from the paper matrix. Therefore, it is necessary to use a stabilizing agent for AuNPs, such as chitosan (Li *et al* 2017). Not only are AuNPs used in paper-based sensors, but also silver nanoparticles (AgNPs) that have been fabricated for ascorbic acid detection (Ferreira *et al* 2015). In this work, the detection of ascorbic acid was achieved via a color change from light yellow to grey due to reaction with AgNPs immobilized on the paper matrix. For quantitative measurement, the color change was captured by a scanner that measures the color intensity value correlated with ascorbic acid concentrations.

3.4.2 Chemiluminescence

Chemiluminescence is an optical detection method that involves light detection produced by a chemical reaction. This chemical reaction occurs when reactive intermediate molecules produce an emission light when returning from their excited state to the ground state. For example, chemiluminescent detection has been developed for uric acid detection based on an enzymatic reaction producing H_2O_2 (Yu *et al* 2011). Here, then H_2O_2 reacted with rhodamine derivatives that have been added to the microfluidic channels of the paper-based biosensor. Another work is based on luminescence when luminol was used. In this case, usually, luminol reacted with iodophenol to produce light as a signal where the peak height correlates with its concentration. This method is one of the simplest optical detection methods as it does not need sophisticated optical instruments. Moreover, it is also one of the most sensitive and precise optical detection methods (Wang *et al* 2012). However, chemiluminescence needs a high enzyme concentration that sometimes could block the paper's porous structure, which in turn could reduce the detection sensitivity.

3.4.3 Electrochemiluminescence

Luminescence or light could also be produced by an electrochemical reaction, in this case, called electrochemiluminescence (Li *et al* 2014). In this approach, a reagent that produces light is usually oxidized at the working electrode with triethylamine (TEA) or H_2O_2 that is obtained by the sample reduction. This redox reaction creates lights or photons resulting in a color change of luminescence. For the quantitative measurement, image evaluation software, such as ImageJ, can be used to measure each color pixel intensity, correlating with the analyte concentrations. For example, an electrochemical luminophore (ruthenium polyvinylpyridine) was activated via oxidation, producing luminescence that has been observed using a CCD camera (Mani *et al* 2013).

3.4.4 Fluorescence

Fluorescence methods are the detection method based on luminescence from fluorophores, as result of interactions with the molecule sample. In this case, to measure the luminescence produced requires fluorometry (Wu *et al* 2018). For example, a paper-based method has been developed based on the fluorescence approach for hydrogen sulfide detection (H_2S) (Petruci and Cardoso 2016). This device contained a light-emitting diode (LED) and a spectrometer, where fluorescein mercury acetate (FMA) was immobilized on the paper surface. The sensing reaction of a paper-based sensor toward H_2S was measured at 470 nm. The results show that response time was 60 s, and the detection limit was found to be 3 ppb. Generally, fluorescent dyes used in this case are biocompatible, however, they should be checked regularly as they can suffer from photobleaching that causes fluorescence intensity loss and loss of reproducibility.

3.5 Water pollution detection

Environmental contaminants, such as heavy metals, pesticides, and toxins are the major elemental contribution in water pollution, and have harmful effects on human health and environment conditions (Schwarzenbach *et al* 2010). Although there are different methods applied for water analysis and monitoring, such as chromatography and spectroscopy methods, assessing the water quality using these methods is time-consuming, costly, and needs sophisticated instruments and skilled personnel (Kuswandi *et al* 2008). Therefore, there is a critical need for simple, rapid, low-cost, and efficient methods for analyzing water pollution. One alternative for water pollution detection is using paper-based optical sensors, both chemical sensors and biosensors (López-Marzo and Merkoçi 2016). In this chapter, the example of the application of paper-based optical sensors using the colorimetric method for the detection of water pollutions is summarized in table 3.3. Here, the target analyte, the sample type, the paper substrate, and the fabrication method, and the limit of detection of the colorimetric paper-based sensor are highlighted.

Regarding the optical detection used in paper-based sensors, the most applied strategies as given in table 3.3 were based on colorimetric detection combined with image processing algorithms using a smartphone camera. For example, colorimetric paper-based sensors have been developed for detection of Hg(II) in different types of water (e.g., tap water, drinking water, river, and pond) (Apilux *et al* 2012, Chen *et al* 2014, 2016a, 2016b, Meelapsom *et al* 2016). The presence of Hg(II) in water at trace levels is often a challenge for its detection and speciation. Moreover, colorimetric paper-based sensors were also developed for the detection of Pb(II) and U(VI) performed by the acquisition of the paper spot images using a digital camera, and ImageJ as color images processing software (Vijitvarasan *et al* 2015, Quesada-González *et al* 2018). Other works on colorimetric paper-based sensors were developed for Fe(III) detection in hot spring waters (Ogawa and Kaneta 2016), and in river and tap waters (Asano and Shiraishi 2015). Several paper-based sensors have also been developed for Cu(II) detection using colorimetric methods (Jayawardane *et al* 2013, Ratnarathorn *et al* 2012, Chaiyo *et al* 2015, 2016). Different approaches were applied in the colorimetric paper-based sensor, such as a polymer inclusion membrane immobilized with a chromogenic reagent (Jayawardane *et al* 2013), AgNPs (Ratnarathorn *et al* 2012), and catalytic etching of silver nanoplates (Chaiyo *et al* 2015, 2016).

Table 3.3. Some examples of the colorimetric paper-based sensors for water pollution detection.

Analyte	Sample type	Paper type	Fabrication	LOD	References
Hg(II)	Drinking and tap water	Filter paper	Wax printing	0.60 μM	Apilux *et al* (2012)
Hg(II)	Pond and river water	Filter paper	Wax printing	50 nM	Chen *et al* (2014)
Hg(II)	Drinking and tap water	Filter paper	Inkjet printing	5 nM	Meelapsom *et al* (2016)
Hg(II)	Pond and river water	Filter paper	Paper cutting	10 nm	Chen *et al* (2016b)
Fe(III)	Groundwater	Filter paper	Wax printing		Dungchai *et al* (2010)
Pb(II)	River water	Chromatography paper	Wax printing	0.3 nM	Vijitvarasan *et al* (2015)
U(VI)	River water	Cellulose and nitrocellulose membranes	Different layers assembling	36 nM	Quesada-González *et al* (2018)
As(III)	Groundwater	Filter paper	Cutting	8 μg L^{-1}	Stocker *et al* (2003)
Cd(II)	Tap water, drinking water	Hi-flow plus nitrocellulose membrane	Different layers assembling	1 ppb	López-Marzo *et al* (2013)
Fe(III)	Hot spring water	Filter paper	Wax printing	0.7 mM	Ogawa and Kaneta (2016)
Fe(III)	River, tap water	Filter paper	Photomask printing	4.0 μM	Asano and Shiraishi (2015)
Cu(II)	Hot tap water, mine tiling water	Filter paper	Inkjet printing	0.94 μM	Jayawardane *et al* (2013)
Cu(II)	Drinking water, groundwater	Filter paper	Wax printing	4.7 nM	Chaiyo *et al* (2015)
Cu(II)	Pond, tap water	Filter paper	Paper cutting	7.8 nM	Ratnarathorn *et al* (2012)
Cu(II)	Stream, groundwater	Filter paper	Wax printing	6.4 nM	Chaiyo *et al* (2016)
Antibiotics	Surface water	Filter paper	Cutting by punching	0.5 μg L^{-1}	Duyen *et al* (2017)
E. coli	Mineral, drinking, aquarium water	Filter paper	Wax printing		Kim and Yeo (2016)

(*Continued*)

Table 3.3. (*Continued*)

Analyte	Sample type	Paper type	Fabrication	LOD	References
E. coli, *Salmonella* spp, *Listeria monocytogenes*	Agricultural water	Chromatography paper	Wax printing	0.1 CFU mL^{-1}	Bisha *et al* (2014)
E. coli, Zika virus	Deionized water	Chromatography paper, grade 2	Wax printing	1 log CFU mL^{-1} 20 pg mL^{-1}	Klug *et al* (2018)
H_2S	Water samples	Filter paper	Paper cutting	0.5 mg L^{-1}	Pla-Tolós *et al* (2016)
Nitrite	Groundwater	Filter paper	Photolithography		Klasner *et al* (2010)
Methomyl	Groundwater	Filter paper	Hot embossing		Yu and Shi (2015)
Organophosphate pesticides	Surface and groundwater	Filter paper	Wax printing	10 nM	Sicard *et al* (2015)

A paper-based colorimetric sensor has been developed for biological contaminants detection in water, such as microorganisms in various water systems (Bisha *et al* 2014, Kim and Yeo 2016). Similar work was also developed for *E. coli* detection, using a smartphone for color image acquisition and processed using ImageJ (Adkins *et al* 2017). Other work based on a smartphone was used as a photometric detector for the detection of the different pathogens (*E. coli* and *Zika virus*) based on a capillary flow rate function (Klug *et al* 2018). This approach used a video recording of an immunoassay, then compared the capillary flows between different concentrations of the target analyte. Another paper-based colorimetric sensor was developed using an LFA for the Cd(II) detection in drinking water (López-Marzo *et al* 2013). The color intensity measurement used a COZART™ RapidScan color reader. Other work using a paper-based colorimetry biosensor was fabricated for semiquantitative detection of As(III) (Stocker *et al* 2003). The biosensor method was based on immobilizing a bacterial on a paper strip. Here, the color change measurement was achieved by spots comparison of standard As (III) concentrations. Moreover, a colorimetric paper-based sensor has been developed for the determination of organophosphate pesticides in water samples (Sicard *et al* 2015).

3.5.1 Heavy metals

Heavy metals (HMs) detection can be performed by colorimetric detection via chemical detection (Kuswandi 2003). HMs discharged from industrial wastewater give a negative impact on the environment when their concentrations exceed a limit. For example, a 3D paper-based sensor was fabricated for colorimetric detection of HMs contaminants in water that allows parallel detection of four HMs (Ni(II), Cd(III), Cr(IV), Cu(II)) (Wang *et al* 2014). The device consisted of four paper layers and three tape layers. In this work, a wax printer was used to create the hydrophobic patterns, while hydrophilic zones were the sample sensing zones that allowed the liquid to flow. The chromogenic reagents specific for the HMs were immobilized in each sensing zone, so they could detect four HMs individually using a smartphone camera to record the color change and subsequently quantify the color intensity. Besides using colorimetric detection, other optical methods could also be applied. For instance, a fluorometric paper-based sensor has been fabricated to monitor HMs (Hg(II), Co(II), Ag(I), and Zn(II)) in wastewater (Feng *et al* 2013), or electrogenerated-chemiluminescence (ECL) detection for Hg(II) and Pb(II) ions (Zhang *et al* 2013).

Another approach in paper-based biosensor development was using biomolecules for HMs detection. For example, a complex biosensor construction consisting of magnetic beads, AuNPs, and the functional nucleotide GR5-DNAzyme was developed for the detection of Pb(II) (Vijitvarasan *et al* 2015). In another approach, a metal-binding protein based on the recombinant human metallothionein 1a was employed for the detection As(III) and Hg(II) in water (Irvine *et al* 2017). Whole-cell living bacteria was also developed as paper-based biosensors using genetically engineered *E. coli*, for detection As(III) (Stocker *et al* 2003).

3.5.2 Other pollutants

Other water pollutants, such as nitrite, and ammonia can be detected using colorimetric detection (Almeida *et al* 2018). Other gas pollutants, such as hydrogen sulfide can also be harmful to human health, even at low concentrations in the blood (Songjaroen *et al* 2012). For this purpose, a colorimetric paper-based sensor has been developed for H_2S detection in environmental water samples (Pla-Tolós *et al* 2016). Here, the sensor is based on the immobilization of *N*,*N*-dimethyl-*p*-phenylenediamine, and $FeCl_3$ onto paper support, where H_2S is adsorbed to produce methylene blue. In other work, a paper-based fluorescence sensor has also been fabricated for H_2S at low concentrations (Petruci and Cardoso 2015). The sensor used filter paper that was impregnated with Pd(II) and ethylene glycol, where Pd(II) was bound to sulfide ligands to form PdS as a sensing mechanism. Moreover, another fluorescence method for H_2S was developed based on the quenching of penicillamine Cu-NC aggregates (Chen *et al* 2016a).

Another method in fluorescence paper-based biosensors has been developed using an immunoassay format and fabricated for ethinylestradiol detection (Scala-Benuzzi *et al* 2018b). Here, an LED was employed as excitation light, while the fluorescence emission was detected using a spectrometer. Whole cells as a biofilm formed from anaerobic sludge were also developed in a paper-based biosensor for the water toxic compounds assessment using a microbial fuel cell device (Chouler *et al* 2018). The detection principle was based on the chemical energy conversion into electricity via the microorganisms' metabolic processes. Here, a microbial biofilm is put on the anode surface, where the electrons' transference to the electrode is mediated by the electroactive bacteria upon their metabolic activity. Therefore any factor disrupting this process, such as water pollution, will reduce the biosensor signal. A similar approach was developed previously for the toxic compound in water using a wastewater bacteria consortium (Xu *et al* 2016).

3.6 Integration with artificial intelligence and machine learning

Currently, there is an increase in problem-solving provided by artificial intelligence (AI) and machine learning (ML) for smart sensor-based systems. For example, real-time AI and ML or new algorithms are needed, and different approaches to integrate these algorithms in a smartphone or other gadgets for optical detection systems of paper-based sensors, to be used for auto sensor calibration, enhancing linear range, etc. A new classification and clustering techniques, learning methods, and data quality methods are needed, including distributed AI algorithms to make a paper-based sensor more accurate in the target analyte detection, in terms of selectivity and sensitivity. For example, the application of an ML algorithm to a luminescence-based sensor has been used for time-resolved luminescence data for the detection of oxygen concentration (Dordević *et al* 2018). The novel method used a neural network (NN) that learns to correlate the quantities sensed of the concentration of oxygen. In this work, the NN training was done with an artificially-created training dataset as the number of measurements is limited. Besides, the temperature that needs to be counted in the luminescence measurements was kept constant but could be predicted by the NN. The best-performing network was optimized by using different architectures and hyperparameter tuning. Then, the trained network was applied to experimental data, to check its generalization when applied to other data. The error in the oxygen concentrations prediction was due to the synthetic training data that were calculated using an approximate model. By using this strategy for the training dataset, the experimental data would reduce the absolute error in the oxygen concentration measurements. This work presented that the implemented approach achieved results comparable with many commercial sensors based on the classical approaches (Wolfbeis 2015, Chu *et al* 2016, Michelucci *et al* 2019). However, using the NN, there is potential to improve its accuracy. Thus, the NN approach in sensor development depends on the analytical model developed that simplifies the data complexity. Therefore, as long as sufficient data are available to train the NN, the system would be able to learn dependence of the oxygen concentration at all parameters provided.

Integration of AI and ML benefits other fields, such as the Internet of Things (IoT), Industry 4.0, etc. Therefore, in paper-based sensor development, these two technologies, would have benefits for simultaneous multiple target analytes, for collecting environmental parameters not only from environmental samples but also in other places, such as homes, offices, vehicles, etc. In this case, the information used to help the decision-making, particularly when it allows the systems to monitor and to learn sensor responses that can be related to target analyte concentration value. Therefore, the future challenges in paper-based sensor development are the innovative works that explore new approaches to applying AI algorithms via smartphone or other gadgets in the measurement system of paper-based sensors. This work will cover new ML models, hybrid AI systems, etc, that could be applied to enhance paper-based sensors' analytical characteristics such as sensitivity and selectivity toward the target analyte.

3.7 Conclusions

The application of paper-based platforms in optical sensor developments has allowed the fabrication of portable analytical devices that are simple, specific, and sensitive as well as user-friendly for the detection of various types of target analytes in water analysis and monitoring. However, most of the reported paper-based sensors were applied to simple water, such as surface water, groundwater, and drinking water, with a small number of organic pollutants compared to wastewater. Therefore, work is needed to develop optical sensors that can be applied in more complex matrices of water samples that need to be taken, involving strategies of sample pretreatment as well as the integration of AI in optical data analysis for smart analysis functions. The most frequently used transducers in optical paper-based sensors comprise colorimetric and fluorescence methods, with open-up analytical strategies in a detection system that can be integrated with AI or ML, related to smart image processing analysis for more accurate measurement, including the possibility for an autocalibration method. Besides the current advances in the development of paper-based optical sensors, the approach for the analysis of various pollutants in the water sample is still open to a new research strategy related with high selectivity and sensitivity for the realization of a smart portable and low-cost analytical tool integrated with AI and ML embedded in an optical detection system for *in situ* water analysis and monitoring.

Acknowledgments

The authors gratefully thank the Ministry of Research and Technology/BRIN, the Republic of Indonesia for supporting this work via the WCR Project 2020–2021.

References

Adkins J A, Boehle K and Friend C *et al* 2017 Colorimetric and electrochemical bacteria detection using printed paper- and transparency-based analytic devices *Anal. Chem.* **89** 3613–21

Ahmed S, Bui M-P N and Abbas A 2016 Paper-based chemical and biological sensors: engineering aspects *Biosens. Bioelectron.* **77** 249–63

Akyazi T, Basabe-Desmonts L and Benito-Lopez F 2018 Review on microfluidic paper-based analytical devices towards commercialisation *Anal. Chim. Acta* **1001** 1–17

Almeida M I G S, Jayawardane B M, Kolev S D and McKelvie I D 2018 Developments of microfluidic paper-based analytical devices (μPADs) for water analysis: a review *Talanta* **177** 176–90

Altundemir S, Uguz A K and Ulgen K 2017 A review on wax printed microfluidic paper-based devices for international health *Biomicrofluidics* **11** 041501

Apilux A, Siangproh W, Praphairaksit N and Chailapakul O 2012 Simple and rapid colorimetric detection of Hg(II) by a paper-based device using silver nanoplates *Talanta* **97** 388–94

Asano H and Shiraishi Y 2015 Development of paper-based microfluidic analytical device for iron assay using photomask printed with 3D printer for fabrication of hydrophilic and hydrophobic zones on paper by photolithography *Anal. Chim. Acta* **883** 55–60

Bisha B, Adkins J A and Jokerst J C *et al* 2014 Colorimetric paper-based detection of *Escherichia coli*, Salmonella spp., and *Listeria monocytogenes* from large volumes of agricultural water *J. Vis. Exp.* **9** 51414

Boehle K E, Carrell C S, Caraway J and Henry C S 2018 Paper-based enzyme competition assay for detecting falsified β-lactam antibiotics *ACS Sens.* **3** 1299–307

Brash J L and Ten Hove P 1993 Protein adsorption studies on 'standard' polymeric materials *J. Biomater. Sci. Polym. Ed.* **4** 591–9

Bruzewicz D A, Reches M and Whitesides G M 2008 Low-cost printing of poly(dimethylsiloxane) barriers to define microchannels in paper *Anal. Chem.* **80** 3387–92

Busa L, Mohammadi S and Maeki M *et al* 2016 Advances in microfluidic paper-based analytical devices for food and water analysis *Micromachines* **7** 86

Carrilho E, Martinez A W and Whitesides G M 2009 Understanding wax printing: a simple micropatterning process for paper-based microfluidics *Anal. Chem.* **81** 7091–5

Chaiyo S, Apiluk A, Siangproh W and Chailapakul O 2016 High sensitivity and specificity simultaneous determination of lead, cadmium and copper using μPAD with dual electrochemical and colorimetric detection *Sens. Actuators B: Chem.* **233** 540–9

Chaiyo S, Siangproh W, Apilux A and Chailapakul O 2015 Highly selective and sensitive paper-based colorimetric sensor using thiosulfate catalytic etching of silver nanoplates for trace determination of copper ions *Anal. Chim. Acta* **866** 75–83

Chen G H, Chen W Y and Yen Y C *et al* 2014 Detection of mercury(II) ions using colorimetric gold nanoparticles on paper-based analytical devices *Anal. Chem.* **86** 6843–9

Chen P C, Li Y C and Ma J Y *et al* 2016a Size-tunable copper nanocluster aggregates and their application in hydrogen sulfide sensing on paper-based devices *Sci. Rep.* **6** 24882

Chen W, Fang X and Li H *et al* 2016b A simple paper-based colorimetric device for rapid mercury (II) assay *Sci. Rep.* **6** 31948

Chitnis G, Ding Z and Chang C L *et al* 2011 Laser-treated hydrophobic paper: an inexpensive microfluidic platform *Lab Chip* **11** 1161–5

Choi D H, Lee S K and Oh Y K *et al* 2010 A dual gold nanoparticle conjugate-based lateral flow assay (LFA) method for the analysis of troponin I *Biosens. Bioelectron.* **25** 1999–2002

Chouler J, Cruz-Izquierdo Á and Rengaraj S *et al* 2018 A screen-printed paper microbial fuel cell biosensor for detection of toxic compounds in water *Biosens. Bioelectron.* **102** 49–56

Christodouleas D C, Nemiroski A, Kumar A A and Whitesides G M 2015 Broadly available imaging devices enable high-quality low-cost photometry *Anal. Chem.* **87** 9170–8

Chu C S, Lin K Z and Tang Y H 2016 A new optical sensor for sensing oxygen based on phase shift detection *Sens. Actuators B: Chem.* **223** 606–12

Chun P 2009 Colloidal gold and other labels for lateral flow immunoassays *Lateral Flow Immunoassay* ed R C Wong and H Y Tse (New York: Humana Press)

Dordević N, Beckwith J S and Yarema M *et al* 2018 Machine learning for analysis of time-resolved luminescence data *ACS Photonics* **5** 4888–95

Dungchai W, Chailapakul O and Henry C S 2009 Electrochemical detection for paper-based microfluidics *Anal. Chem.* **81** 5821–6

Dungchai W, Chailapakul O and Henry C S 2010 Use of multiple colorimetric indicators for paper-based microfluidic devices *Anal. Chim. Acta* **674** 227–33

Dungchai W, Chailapakul O and Henry C S 2011 A low-cost, simple, and rapid fabrication method for paper-based microfluidics using wax screen-printing *Analyst* **136** 77–82

Duyen T T M, Matsuura H and Ujiie K *et al* 2017 Paper-based colorimetric biosensor for antibiotics inhibiting bacterial protein synthesis *J. Biosci. Bioeng.* **123** 96–100

Evans A E, Mateo-Sagasta J and Qadir M *et al* 2019 Agricultural water pollution: key knowledge gaps and research needs *Curr. Opin. Environ. Sustain.* **36** 20–7

Feng L, Li H and Niu L Y *et al* 2013 A fluorometric paper-based sensor array for the discrimination of heavy-metal ions *Talanta* **108** 103–8

Ferreira D C M, Giordano G F and Soares C C D S P *et al* 2015 Optical paper-based sensor for ascorbic acid quantification using silver nanoparticles *Talanta* **141** 188–94

Gaspar C, Olkkonen J, Passoja S and Smolander M 2017 Paper as active layer in inkjet-printed capacitive humidity sensors *Sensors (Switzerland)* **17** 1464

Gervais L and Delamarche E 2009 Toward one-step point-of-care immunodiagnostics using capillary-driven microfluidics and PDMS substrates *Lab Chip* **9** 3330–7

Gosselin D, Belgacem M N and Joyard-Pitiot B *et al* 2017 Low-cost embossed-paper micro-channels for spontaneous capillary flow *Sens. Actuators B: Chem.* **248** 395–401

Hashemi Goradel N, Mirzaei H and Sahebkar A *et al* 2018 Biosensors for the detection of environmental and urban pollutions *J. Cell. Biochem.* **119** 207–12

Hidayat M A, Illa R and Indah C *et al* 2019 The CUPRAC—paper microzone plates as a simple and rapid method for total antioxidant capacity determination of plant extract *Eur. Food Res. Technol.* **245** 2063–70

Hossain S M Z, Luckham R E, McFadden M J and Brennan J D 2009 Reagentless bidirectional lateral flow bioactive paper sensors for detection of pesticides in beverage and food samples *Anal. Chem.* **81** 9055–64

Hosseini S, Vázquez-Villegas P and Martínez-Chapa S O 2017 Paper and fiber-based bio-diagnostic platforms: current challenges and future needs *Appl. Sci.* **7** 863

Irvine G, Tan S and Stillman M 2017 A simple metallothionein-based biosensor for enhanced detection of arsenic and mercury *Biosensors* **7** 14

Jayawardane B M, Coo L D L C, Cattrall R W and Kolev S D 2013 The use of a polymer inclusion membrane in a paper-based sensor for the selective determination of Cu(II) *Anal. Chim. Acta* **803** 106–12

Jéquier E and Constant F 2010 Water as an essential nutrient: the physiological basis of hydration *Eur. J. Clin. Nutr.* **64** 115–23

Jokerst J C, Adkins J A and Bisha B *et al* 2012 Development of a paper-based analytical device for colorimetric detection of select foodborne pathogens *Anal. Chem.* **84** 2900–7

Khan M S, Fon D and Li X *et al* 2010 Biosurface engineering through ink jet printing *Colloids Surf. B: Biointerfaces* **75** 441–7

Kim J Y and Yeo M K 2016 A fabricated microfluidic paper-based analytical device (μPAD) for *in situ* rapid colorimetric detection of microorganisms in environmental water samples *Mol. Cell. Toxicol.* **12** 101–9

Kim Y S, Yang Y and Henry C S 2018 Laminated and infused Parafilm®—paper for paper-based analytical devices *Sens. Actuators B: Chem.* **255** 3654–61

Klasner S A, Price A K and Hoeman K W *et al* 2010 Paper-based microfluidic devices for analysis of clinically relevant analytes present in urine and saliva *Anal. Bioanal. Chem.* **397** 1821–9

Klug K E, Reynolds K A and Yoon J-Y 2018 A capillary flow dynamics-based sensing modality for direct environmental pathogen monitoring *Chem. A Eur. J.* **24** 6025–9

Koop S H A and Van Leeuwen C 2017 The challenges of water, waste and climate change in cities *Environ. Dev. Sustain.* **19** 285–418

Kung C-T, Hou C-Y, Wang Y-N and Fu L-M 2019 Microfluidic paper-based analytical devices for environmental analysis of soil, air, ecology and river water *Sens. Actuators B: Chem.* 126855

Kuswandi B 2003 Simple optical fibre biosensor based on immobilised enzyme for monitoring of trace heavy metal ions *Anal. Bioanal. Chem.* **376** 1104–10

Kuswandi B, Fikriyah C I and Gani A A 2008 An optical fiber biosensor for chlorpyrifos using a single sol-gel film containing acetylcholinesterase and bromothymol blue *Talanta* **74** 613–8

Kuswandi B, Irmawati T and Hidayat M A *et al* 2014 A simple visual ethanol biosensor based on alcohol oxidase immobilized onto polyaniline film for halal verification of fermented beverage samples *Sensors (Basel)* **14** 2135–49

Kuswandi B, Nuriman and Dam H H *et al* 2007a Development of a disposable mercury ion-selective optode based on trityl-picolinamide as ionophore *Anal. Chim. Acta* **591** 208–13

Kuswandi B, Nuriman, Huskens J and Verboom W 2007b Optical sensing systems for microfluidic devices: a review *Anal. Chim. Acta* **601** 141–55

Kuswandi B, Taib M N and Narayanaswamy R 1999 New solid state instrument for optical toxic metal ions sensing *Sens. Actuators A: Phys.* **76** 183–90

Kuswandi B, Tombelli S, Marazza G and Mascini M 2005 Recent advances in optical DNA biosensors technology *Chimia (Aarau)* **59** 236–42

Li J, Wang X and Huo D *et al* 2017 Colorimetric measurement of Fe^{3+} using a functional paper-based sensor based on catalytic oxidation of gold nanoparticles *Sens. Actuators B: Chem.* **242** 1265–71

Li L, Li W and Ma C *et al* 2014 Paper-based electrochemiluminescence immunodevice for carcinoembryonic antigen using nanoporous gold-chitosan hybrids and graphene quantum dots functionalized Au@Pt *Sens. Actuators B: Chem.* **202** 314–22

Liana D D, Raguse B, Justin Gooding J and Chow E 2012 Recent advances in paper-based sensors *Sensors (Switzerland)* **12** 11505–26

Lisa M, Chouhan R S and Vinayaka A C *et al* 2009 Gold nanoparticles based dipstick immunoassay for the rapid detection of dichlorodiphenyltrichloroethane: an organochlorine pesticide *Biosens. Bioelectron.* **25** 224–7

López-Marzo A M and Merkoçi A 2016 Paper-based sensors and assays: a success of the engineering design and the convergence of knowledge areas *Lab Chip* **16** 3150–76

López-Marzo A M, Pons J, Blake D A and Merkoçi A 2013 High sensitive gold-nanoparticle based lateral flow Immunodevice for Cd^{2+} detection in drinking waters *Biosens. Bioelectron.* **47** 190–8

Mahmud M, Blondeel E, Kaddoura M and MacDonald B 2018 Features in microfluidic paper-based devices made by laser cutting: how small can they be? *Micromachines* **9** 220

Mani V, Kadimisetty K and Malla S *et al* 2013 Paper-based electrochemiluminescent screening for genotoxic activity in the environment *Environ. Sci. Technol.* **47** 1937–44

Martinez A W, Phillips S T and Carrilho E *et al* 2008 Simple telemedicine for developing regions: camera phones and paper-based microfluidic devices for real-time, off-site diagnosis *Anal. Chem.* **80** 3699–707

McCracken K E, Tat T and Paz V *et al* 2017 Immunoagglutinated particle rheology sensing on a microfluidic paper-based analytical device for pathogen detection *2017 ASABE Annual Int. Meeting. American Society of Agricultural and Biological Engineers*

Meelapsom R, Jarujamrus P and Amatatongchai M *et al* 2016 Chromatic analysis by monitoring unmodified silver nanoparticles reduction on double layer microfluidic paper-based analytical devices for selective and sensitive determination of mercury(II) *Talanta* **155** 193–201

Michelucci U, Baumgartner M and Venturini F 2019 Optical oxygen sensing with artificial intelligence *Sensors* **19** 777

Nilghaz A, Liu X and Ma L *et al* 2019 Development of fabric-based microfluidic devices by wax printing *Cellulose* **26** 3589–99

Ogawa K and Kaneta T 2016 Determination of iron ion in the water of a natural hot spring using microfluidic paper-based analytical devices *Anal. Sci.* **32** 31–4

Ouyang L, Wang C and Du F *et al* 2014 Electrochromatographic separations of multi-component metal complexes on a microfluidic paper-based device with a simplified photolithography *RSC Adv.* **4** 1093–101

Pan M, Ma T, Yang J, Li S, Liu S and Wang S 2020 Development of lateral flow immunochromatographic assays using colloidal Au sphere and nanorods as signal marker for the determination of zearalenone in cereals *Foods 2020* **9** 281

Petruci J F D S and Cardoso A A 2016 Portable and disposable paper-based fluorescent sensor for *in situ* gaseous hydrogen sulfide determination in near real-time *Anal. Chem.* **88** 11714–9

Petruci J F D S and Cardoso A A 2015 Sensitive luminescent paper-based sensor for the determination of gaseous hydrogen sulfide *Anal. Methods* **7** 2687–92

Pla-Tolós J, Moliner-Martínez Y and Verdú-Andrés J *et al* 2016 New optical paper sensor for *in situ* measurement of hydrogen sulphide in waters and atmospheres *Talanta* **156–57** 79–86

Posthuma-Trumpie G A, Korf J and Van Amerongen A 2009 Lateral flow (immuno)assay: its strengths, weaknesses, opportunities and threats. A literature survey *Anal. Bioanal. Chem.* **393** 569–82

Quesada-González D, Jairo G A and Blake R C *et al* 2018 Uranium (VI) detection in groundwater using a gold nanoparticle/paper-based lateral flow device *Sci. Rep.* **8** 8–15

Ratnarathorn N, Chailapakul O, Henry C S and Dungchai W 2012 Simple silver nanoparticle colorimetric sensing for copper by paper-based devices *Talanta* **99** 552–7

Rengaraj S, Cruz-Izquierdo Á, Scott J L and Di Lorenzo M 2018 Impedimetric paper-based biosensor for the detection of bacterial contamination in water *Sens. Actuators B: Chem.* **265** 50–8

Santhiago M, Nery E W, Santos G P and Kubota L T 2014 Microfluidic paper-based devices for bioanalytical applications *Bioanalysis* **6** 89–106

Scala-Benuzzi M L, Raba J and Soler-Illia G J A A *et al* 2018a Novel electrochemical paper-based immunocapture assay for the quantitative determination of ethinylestradiol in water samples *Anal. Chem.* **90** 4104–11

Scala-Benuzzi M L, Takara E A and Alderete M *et al* 2018b Ethinylestradiol quantification in drinking water sources using a fluorescent paper based immunosensor *Microchem. J.* **141** 287–93

Schwarzenbach R P, Egli T and Hofstetter T B *et al* 2010 Global water pollution and human health *Annu. Rev. Environ. Resour.* **35** 109–36

Shangguan J W, Liu Y and Bin P J *et al* 2017 Microfluidic PDMS on paper (POP) devices *Lab Chip* **17** 120–7

Sher M, Zhuang R, Demirci U and Asghar W 2017 Paper-based analytical devices for clinical diagnosis: recent advances in the fabrication techniques and sensing mechanisms *Expert Rev. Mol. Diagn.* **17** 351–66

Sicard C, Glen C and Aubie B *et al* 2015 Tools for water quality monitoring and mapping using paper-based sensors and cell phones *Water Res.* **70** 360–9

Singh A T *et al* 2018 Paper-based sensors: emerging themes and applications *Sensors (Switzerland)* **18** 1–22

Songjaroen T, Dungchai W and Chailapakul O *et al* 2012 Blood separation on microfluidic paper-based analytical devices *Lab Chip* **12** 3392

Spicar-Mihalic P, Toley B and Houghtaling J *et al* 2013 CO_2 laser cutting and ablative etching for the fabrication of paper-based devices *J. Micromech. Microeng.* **23** 067003

Stocker J, Balluch D and Gsell M *et al* 2003 Development of a set of simple bacterial biosensors for quantitative and rapid measurements of arsenite and arsenate in potable water *Environ. Sci. Technol.* **37** 4743–50

Tsai T T, Huang C Y and Chen C A *et al* 2017 Diagnosis of tuberculosis using colorimetric gold nanoparticles on a paper-based analytical device *ACS Sens.* **2** 1345–54

Tyagi C, Tomar L K and Singh H 2009 Surface modification of cellulose filter paper by glycidyl methacrylate grafting for biomolecule immobilization: influence of grafting parameters and urease immobilization *J. Appl. Polym. Sci.* **111** 1381–90

Vijitvarasan P, Oaew S and Surareungchai W 2015 Paper-based scanometric assay for lead ion detection using DNAzyme *Anal. Chim. Acta* **896** 152–9

Wang H, Li Y J and Wei J F *et al* 2014 Paper-based three-dimensional microfluidic device for monitoring of heavy metals with a camera cell phone *Anal. Bioanal. Chem.* **406** 2799–807
Wang S, Ge L and Song X *et al* 2012 Paper-based chemiluminescence ELISA: lab-on-paper based on chitosan modified paper device and wax-screen-printing *Biosens. Bioelectron.* **31** 212–8
Wolfbeis O S 2015 Luminescent sensing and imaging of oxygen: fierce competition to the Clark electrode *BioEssays* **37** 921–8
Wu M, Lai Q and Ju Q *et al* 2018 Paper-based fluorogenic devices for *in vitro* diagnostics *Biosens. Bioelectron.* **102** 256–66
Xie Y-J, Yang Y and Kong W-J *et al* 2015 Application of nanoparticle probe-based lateral flow immunochromatographic assay in mycotoxins detection *Chinese J. Anal. Chem.* **43** 618–28
Xu Z, Liu Y and Williams I *et al* 2016 Disposable self-support paper-based multi-anode microbial fuel cell (PMMFC) integrated with power management system (PMS) as the real time 'shock' biosensor for wastewater *Biosens. Bioelectron.* **85** 232–9
Yetisen A, Akram M and CR L 2013 Paper-based microfluidic point-of-care diagnostic devices *Lab Chip* **13** 2210–51
Yetisen A K, Martinez-Hurtado J L and Garcia-Melendrez A *et al* 2014 A smartphone algorithm with inter-phone repeatability for the analysis of colorimetric tests *Sens. Actuators B: Chem.* **196** 156–60
Yu J, Wang S, Ge L and Ge S 2011 A novel chemiluminescence paper microfluidic biosensor based on enzymatic reaction for uric acid determination *Biosens. Bioelectron.* **26** 3284–9
Yu L and Shi Z Z 2015 Microfluidic paper-based analytical devices fabricated by low-cost photolithography and embossing of Parafilm® *Lab Chip* **15** 1642–5
Zhang M, Ge L and Ge S *et al* 2013 Three-dimensional paper-based electrochemiluminescence device for simultaneous detection of Pb^{2+} and Hg^{2+} based on potential-control technique *Biosens. Bioelectron.* **41** 544–50

Chapter 4

Nanocomposite materials for water purification: synthesis, characterization, and applications

Naumih M Noah and Peter M Ndangili

Water is one of the most precious natural resources whose quality and quantity are essential for the survival of all living things. The United Nations recognize reliable access to clean, affordable, and sufficient water as one of the most basic humanitarian goals in the 21st century. However, this has been a major global challenge to date due to degrading water quality from the release of large amounts of pollutants into water bodies. Currently, an estimated 1.2 billion people worldwide have inadequate access to fresh potable drinking water while 2.5 billion lack access to sanitation. For this reason, the treatment of wastewater is essential in order to satisfy the growing demands for sufficient clean water for human and industrial use. To achieve affordable and sustainable water treatment techniques, appropriate technologies that give rise to low-cost materials with high pollutant separation capacity, porosity, and reusability are required. Nanotechnology has huge potential for addressing these challenges since it presents an opportunity to develop a wide variety of materials for water purification by modulating nanomaterial's properties such as mechanical strength, dispersibility, porosity, hydrophobicity, and hydrophilicity among others. Consequently, various nanomaterials have been developed and applied in water treatment. Researchers have, however, been met with nanoparticle agglomeration, an aspect that limits their use in water treatment. To address this challenge, nanocomposites have been developed by the combination of more than two materials with various physical and chemical properties and a unique interface. This chapter discussed the various types of nanocomposites developed for water treatment. Specifically, it focuses on polymeric nanocomposites and non-polymeric nanocomposites. The aspects discussed for each of these materials are their synthesis, characterization as well as specific examples.

doi:10.1088/978-0-7503-3047-3ch4

4.1 Introduction

Water is an important natural resource of life and one of the most vital bases for the living system, and is used in daily life activities for human survival and development [1–4]. Although 71% of the Earth's surface is covered with freshwater resources that can be used for human consumption, we are still facing various problems due to a lack of clean and fresh water [2, 5]. The World Health Organization (WHO) 2014 report on freshwater supply and sanitation estimated that 1.2 billion people have inadequate access to fresh portable drinking water, and another 2.5 billion people were without access to sanitation [5, 6]. This has been attributed to the rapid development of industries and increasing human activities [4, 7, 8], which seriously endanger the freshwater resource and ecological environment, hence undermining their quality [3]. Diseases from unsafe drinking water and lack of basic sanitation kill more people every year than all forms of violence, accidents, and other diseases. Globally, as reported by the WHO, 1 in 8 people cannot access safe drinking water and about 3.5 million people, of which 84% are children and 98% occur in the developing countries, die every year from water-borne diseases emanating from drinking unsafe water [5, 6, 9, 10]. This is an indication that water pollution by various pollutants is becoming an alarming global environmental problem [2].

Water pollution normally emanates from increased anthropogenic activities such as processing, human and domestic waste, agriculture, mining, manufacturing industries, and pharmaceutical and petroleum industries, which are closely dependent on the demands of an increasing population and industrialization [11, 12]. These untreated toxic wastes as effluents or leachates find their way into drinking water systems, and therefore water treatment at the point of use and pollutant source would greatly reduce infant deaths and hospital visits as a result of unsafe drinking water [11]. The need to provide safe drinking water especially to poor people in developing countries cannot be overemphasized [13].

The conventional water purification methods to provide clean and safe drinking water to the public include coagulation, sedimentation, filtration, and disinfection with a combination of coagulation, sedimentation, and filtration as the most widely applied water treatment technology worldwide, having been used since the early 20th century [14]. Coagulation can successfully remove a large number of organic compounds, including some dissolved organic material or dissolved organic carbon (DOC), as well as suspended particles, including inorganic precipitates such as iron. However, while coagulation and filtration can remove particles and some dissolved matter, the water may still contain pathogens such as viruses and bacteria and hence it cannot produce safe drinking water [14]. With the increasing global need for water and wastewater treatment, there has been widespread development of large-scale membrane filtration processes [15, 16]. Compared to the conventional separation processes, membranes are known to offer flexible economical alternatives based on their usage, their high capacity for removal, their availability, and low energy consumption [17, 18]. The most commonly used membranes have been the synthetic ones constructed from organic or polymeric and inorganic or ceramic materials, with the polymeric membranes attracting the most interest due to their mechanical strength and chemical stability [17, 19]. However, there has been an unavoidable occurrence of filtration membrane fouling throughout the water treatment industry [15, 17]. The membrane fouling arises when microbes, suspended solids, salts, and organic material are deposited on the surface of membranes during selective water permeation [15, 17]. This normally reduces permeation flux, and affects selectivity and separability when the filter is working [17]. Membrane fouling can become irreversible/permanent when the foulant is chemisorbed by the surface and/or the pores of the membrane, causing flux decline hence shortening the lifespan of the membrane. To remove the membranes, back-washing and cleaning has to be used and can cause operational disruption as processes have to be shut down for maintenance as well as causing problems regarding the disposal of waste [17, 20, 21]. Recently, increasing water demands have forced research into membranes with a superior water flux. With this progress in water flux enhancement, research into membrane anti-fouling has become more and more pressing as membrane fouling worsens along with increasing water flux [22, 23]. Fouled membranes have substantially shorter lifespan and lessened filtration efficiency. The fouling problem is most pronounced in membranes with decreasing pore size. Reverse osmosis (RO) and forward osmosis (FO) membranes (0.1–1 nm pore size) depict the highest fouling contrasted to nano, ultra and micro filtration membranes. These membranes also require increasingly high pressure operations to boost water flux and are expensive to maintain and operate [24, 25].

Nanotechnology, the science and art of manipulating matter at the atomic and molecular level, has the potential to deliver affordable and effective solutions for this problem in water sanitation/purification, providing access to safe drinking water to millions of people [26] and therefore contributing to poverty alleviation and achievement of the sustainable development goals (SDGs). Due to its rapid advancement, nanotechnology, and the production of advanced nanomaterials offer significant opportunities for an extensive range of applications for the detection and remediation of a broad range of environmental contaminants [27]. The integration of polymeric materials with synthetic nanomaterials may permit unprecedented ability to detect, disinfect and completely remove pathogens in water [28]. Nanoparticle insertion onto polymeric materials continues to appeal as a suitable process since it leads to versatile materials of diverse applications [27, 29] since they have a large surface-to-volume ratio, thus surface-related phenomena and properties are drastically affected with a slight modification of size, shape and surrounding media. Consequently, the desired electronic, catalytic, and optical properties of nanoparticles, depending on the application, can be tuned by generating nanoparticles of defined size and shape in selected media allowing the development of new effective nanomaterials and nanodevices [27, 29–31]. Polymeric materials provide an excellent dispersion and sequestering environment for nanoparticles [32]. Water and waste-water treatment in particular has harnessed the potential of nano-assisted polymeric materials with the desired antifouling, hydrophilicity, photocatalytic, self-cleaning and photo degrading properties [10, 17, 21, 33]. Surface modification of polyaniline, polysulfone, cellulose acetate among others has been done with nanoparticles to make better membrane materials [34]. Most of these polymers have the character to act as reductants and stabilizers of metal nanoparticles [35]. However, the application of them is limited due to low processability, insolubility in a wide range of solvents, and non-biodegradability [35]. Polyamic acid (PAA), the highly reactive intermediate of polyimides, is soluble in both organic and aqueous solvents, demonstrates good mechanical stability, and acts as a better redox polymer than previously employed polymeric materials [36]. In this book chapter, we will present various Nanocomposites focusing on the preparation, characterization, and functionalization of nanomaterials on membranes and their potential application in water purification.

4.2 Nanocomposites for water filtration

Nanocomposites are multi-phasic materials, formed through the combination of more than two materials with various physical and chemical properties and unique interface [1, 4, 37, 38]. They are reported as multiple phase materials of the 21st century in which one of the constituents is a matrix (major phase) while the other one is a reinforcement (minor phase) [39] which shows dimensions in the nano range (10–100 nm) [1, 40–42]. These nanocomposite materials have emerged as suitable alternatives to overcome limitations of different engineering materials and have been applied in various fields such as life sciences, drug distribution schemes, and drinking and wastewater treatment in one-dimensional, two-dimensional, and three-dimensional form. They also include amorphous materials of dissimilar components, mixed at a nanometer scale. In nanocomposites, the nanomaterials are usually merged within diverse functionalized materials like multi-walled carbon nanotube (MWCNT), activated carbon, graphene oxide, or polymeric materials, and hence they are classified either as non-polymeric and polymeric nanocomposites based on their dispersed matrix and dispersed phase materials, as illustrated in figure 4.1 [4]. They are further classified into ceramic matrix nanocomposites, metal matrix nanocomposites, carbon–carbon nanocomposites, hybrid matrix nanocomposites and polymer matrix

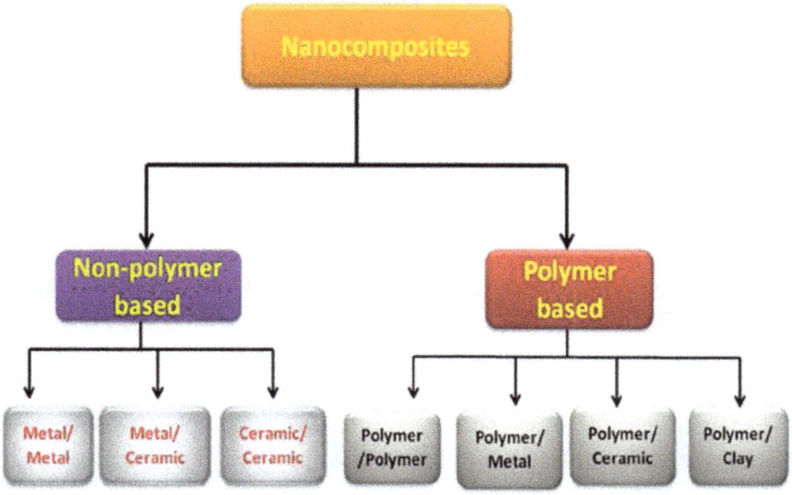

Figure 4.1. Classification of nanocomposites. Reproduced from [4], CC BY 4.0.

nanocomposites [39, 41, 43]. This classification depends on the matrix materials used in their synthesis. The non-polymeric nanocomposites such as ceramic matrix nanocomposite mainly consist of either aluminum oxide (Al_2O_3), zirconia (ZrO_2) or silicon carbide (SiC) [44], which have been reported to have higher thermal, hydrophilicity, and surface charge, mechanical and chemical stability than the polymeric nanocomposites and hence can be used under extreme pH and temperature conditions and in a high oxidizing environment [16, 45]. The polymeric nanocomposites on the other hand offer great design flexibility and are generally cheaper [16]. They constitute an inorganic material conjugated into a polymeric matrix [46]. Polymer matrix nanocomposites are made up of a polymeric or copolymeric material in which nanoparticles or nanofillers are dispersed throughout the polymer chain. These nanocomposite materials have attracted tremendous growth rate in the recent past, due to the various ways in which they can be functionalized, consequently permitting multi-applications [47, 48]. They have found applications in water purification, whose main aim is to provide quality water to human and animal consumption. Water purification involves filtration of surface water, wastewater as well as groundwater [49]. A number of processes such as microbial disinfection, photocatalysis as well as adsorption/desorption take place in water filtration. The importance of nanocomposites in water purification is attributed to their ability to provide active surface sites for adsorption of metal contaminants, reduced agglomeration, and antimicrobial activities [50–52]. The application of nanocomposites in the purification of water through microbial disinfection, photocatalysis, as well as adsorption/desorption, is met with leaching challenges [50]. For this reason, research has focused on filtration as an effective alternative method of water purification. This is because, water filtration processes can be designed such that the filters contain materials that enable microbial disinfection, photocatalysis as well as adsorption/desorption; thereby achieving water quality by a single process. Consequently, a lot of research activities have been done to develop nanofilters for water. In the following section, we discuss the synthesis, characterization and potential application of both the non-polymeric and the polymeric nanocomposites in water purification.

4.2.1 Non-polymeric nanocomposites

Metal nanoparticles, metal oxide nanoparticles, and metal nanocomposites possess characteristic properties that make them attractive adsorbents in water purification. They have large surface areas compared to bulk materials and are capable of being functionalized with a wide variety of functional groups to significantly increase their affinity toward target compounds that pollute water [53]. Metal nanocomposites find applications in techniques such as adsorption, precipitation, ion exchange, reverse osmosis, electrochemical treatments, membrane filtration, evaporation, flotation, oxidation, and bio-sorption, all of which give rise to environmental detoxification of water [54]. The use of metal nanocomposites in adsorption has been found to efficiently remove toxic metal ions and bacterial pathogens from water [55]. One of the concepts of water purification involves the oxidation of water molecules using solar energy to produce oxidative radicals that kill microorganisms in water [56].

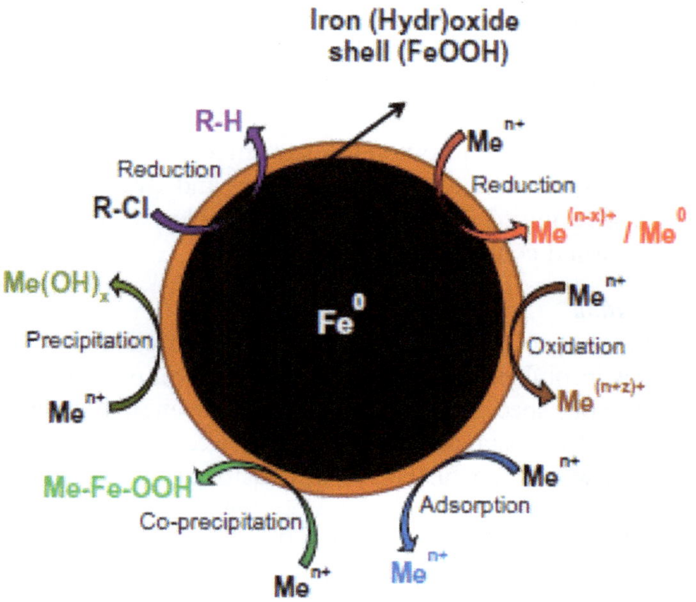

Figure 4.2. Core–shell structure of nanoscale zero-valent iron (nZVI) depicting various mechanisms for the removal of metals and chlorinated compounds. Adapted with permission from [59], Copyright 2013 Elsevier.

This concept requires metal nanocomposites with high photocatalytic activity such as those containing titanium dioxide, iron oxide, nickel oxide, zinc oxide, tungsten oxide, copper oxide, and aluminum oxide [50, 56, 57]. Another concept of water purification involves selective and specific removal of heavy metal water pollutants such as mercury, antimony, arsenic, chromium, lead, nickel, and cadmium using appropriate adsorbents [58]. The metal nanocomposites required for this concept should have high affinity to the specific metals targeted for removal and high surface area to ensure availability adequate adsorption sites and increase adsorption efficiency [58]. Adsorption presents an effective and selective method of removing these metal pollutants, especially if they are dissolved in drinking water. Figure 4.2 illustrates the mechanisms for the removal of pollutants from water using metal nanocomposites.

These heavy metals may also be removed from water by reverse osmosis, ion exchange, air stripping, or chemical precipitation.

4.2.1.1 Synthesis

Several techniques are used in the synthesis of metal nanocomposites. These techniques can be classified into liquid phase, gas phase, and vapor phase [50]. The main synthesis methods include electrochemical, precipitation, solvothermal, thermal decomposition, ball milling, green synthesis, and vapor condensation [58]. Solvothermal, thermal decomposition, and vapor condensation are high energy demanding methods, require special facilities, and are consequently costly. Electrochemical, precipitation and ball mining are characterized by high production rate, water compatibility, low or medium cost, and ecofriendliness.

4.2.1.1.1 Electrochemical synthesis

Nanocomposites synthesized by this method are electrodeposited on the surface of a suitable substrate using one of the many electrochemical techniques such as electrolysis. This method has been used for the synthesis of nanocomposites such as ZnO–WO_3 [60] whose water filtration mechanism involves photolytic degradation of water pollutants such as detergents, pesticides, and dyes [60]. This method is scalable and can give mixed metal oxide nanoparticles with high photocatalytic properties under normal conditions.

4.2.1.1.2 Precipitation

Precipitation synthesis processes involve ion exchange reactions in which one of the products formed is insoluble. These reactions are not complex and mostly proceed spontaneously since one species in the reaction is strong enough to displace the other from its compound. For instance, the reaction between NaOH and $ZnSO_4$ forms insoluble $Zn(OH)_2$, which can further be heated to give ZnO nanoparticles. This method has been used to prepare both doped and undoped ZnO nanoparticles such as ZnO [61], $ZnWO_4$ [62], Ce-doped ZnO [63], and many other different types of metal oxide nanoparticles. Magnetic metal particles are known to aggregate to form bulk materials, which are thermodynamically stable [64]. This aggregation decreases their surface area and hence the adsorption sites increase. Synthesis procedures for these particles therefore should include steps that ensure post-synthesis non-aggregation. This can be done using surface modification techniques such as coating the nanoparticle with a more stable shell; for instance, a metal oxide layer. This is because metal oxide particles once hydrated form hydroxyl groups. These hydroxyl groups can undergo protonation and deprotonation reactions depending on the pH to form negative charges or positive charges on the surface of the nanoparticle. A positive charge density is developed at pH values less than the point of zero charges (PZC) pH, whereas a negative charge density develops at pH values higher than PZC pH [64]. A dense negative charge on a nanomaterial surface creates electrostatic repulsive forces on the individual particles, hence keeping them apart. The same principle applies for a dense positive charge on the nanoparticle surface. For removal of heavy metal cation pollutants in water, a negative charge density on the nanoparticle surface is desirable since it would electrostatically attract the positive metal ions, while at the same time ensuring non aggregation, as illustrated in figure 4.3.

Co-precipitation is also possible toward the synthesis of metal nanocomposites. It has been applied for the synthesis of metal nanocomposites such as MnFe2O4@Mn–Co oxide core–shell nanoparticles (co-precipitation) [64], magnetite–silica [66]. Co-precipitation is said to be a simple and highly scalable synthesis method, which gives nanosized super magnetic metal nanoparticles such as iron oxide. These nanoparticles however have poor crystallinity due to agglomeration (aggregation). This can be addressed through surface modification as discussed above. Co-precipitation makes it possible to achieve a synthesis of the core material as well as the surface modification in one step [66]. This is demonstrated by the successful synthesis of magnetite–silica using iron(II) and (III) salts by an alkaline

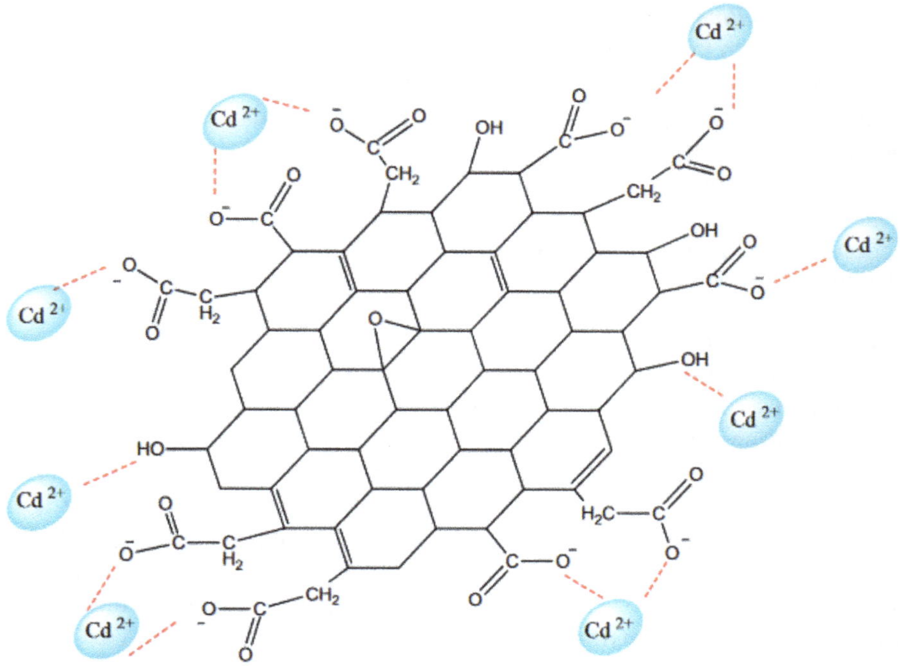

Figure 4.3. Possible adsorption mechanism of Cd^{2+} onto $NiFe_2O_4$/HAP/GQDs surface. Reproduced from [65].

solution of sodium silicate [66]. The use of silica in this method ensures the stability of the synthesized material in buffers, allows for the multi-composite formation, and is biocompatible. It also increases the surface area of the nanocomposite, thereby enabling the entrapment of metal pollutants. Sodium silicate is a good source of silica in this method and can be obtained commercially or extracted from ash of rice husks.

4.2.1.1.3 Solvothermal synthesis

In solvothermal synthesis, a chemical reaction takes place in an organic solvent at a temperature above its normal boiling point in which the reaction mixture is contained in a sealed container such as an autoclave [67]. Some nanocomposites synthesized by this method and applied in water treatment include Al-doped ZnO [68].

4.2.1.1.4 Thermal decomposition synthesis

This method is highly studied for the synthesis of iron oxide nanoparticles [69]. It involves temperature-driven breakdown of organometallic precursors mixed with stabilizers and in the presence or absence of oxygen to give metal oxides. The organometallic precursors are prepared by mixing suitable salts of the metal with a soluble organic salt. Common organometallic precursors used for the synthesis of iron oxide nanoparticles are iron oleate [69–72], iron acetate [73], and iron acetylacetonate [71]. In the absence of oxygen, the nanoparticles formed are characterized by crystalline defects, poor magnetic properties, and a mixture of oxide phases [74].

4.2.1.1.5 Ball milling synthesis

This method is also known as alloying and has been used for the fabrication of many surface-modified nanocomposites such as CuO modified biochar [75] biochar/Fe_3O_4, activated carbon/Fe_3O_4 [76], ZnO@sulfur/carbon nanotubes, and $Ni(OH)_2$ sulfur/carbon nanotubes composites [77]. These reactions can take place under solvent-free conditions [78]. For applications in water filtration via adsorption, ball mining synthesis of nanocomposites is a desirable method since it allows for easy particle size reduction, accompanied by increased specific surface area. This leads to enhanced availability of adsorption sites and consequently improved adsorption capacity.

4.2.1.1.6 Green synthesis

Green synthesis normally uses reducing agents obtained from plant extracts to reduce aqueous species of the metal in solution [79]. For example, various plant extracts including *Clitoria ternatea,* plants (green tea (*Camellia sinensis*), alfalfa (*Medicago sativa*), lemongrass (*Cymbopogon flexuosus*), geranium (*Pelargonium graveolens*)) [80] Citrus paradise (Grapefruit red) [81], and *Solanum nigrum* [82] have been used as reducing agents to reduce Ag^+ from an aqueous solution of $AgNO_3$ [79]. Also, biological agents such as bacteria, fungi, and algae have been used as reducing agents for silver [79, 80]. These agents present cost-effective and environmentally friendly pathways for nanoparticle synthesis [79]. Various nanoparticles such as silver nanoparticles, magnesium oxide nanoparticles exhibit antimicrobial properties [79, 83] making them suitable for use in water filtration [84]. The synthesis of silver nanoparticles for water filtration is designed in such a way that they are stabilized through surface engineering, impregnated with other materials placed on substrates [85] such as sand, zeolites [86], fiberglass [87], activated carbon [88], blotter paper [89], cellulose filters [90], ceramic filters [91], or polyurethane foams [92] among others [79]. A design of a silver nanoparticle filtering bottle cap described by [80] is shown in figure 4.4. In this design, a double-threaded bottle cap is used to allow the flow of contaminated water from the support

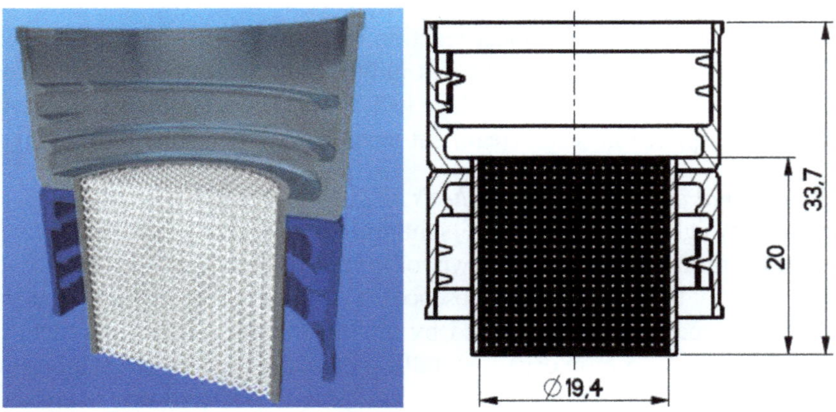

Figure 4.4. A design of a silver nanoparticle filtering bottle cap. Reproduced from [80]. Copyright the authors, published by EDP Sciences. CC BY 4.0.

bottle to the receiver bottle. The bottle cap is fitted with a 19.4 mm perforated cylindrical filter impregnated with silver nanoparticles.

4.2.1.1.7 Vapor condensation

Vapor condensation, also known as vapor deposition may take place physically or chemically. In physical vapor deposition, materials are formed by condensation of atoms or molecules on a substrate surface as a result of evaporation, sputtering, or ion bombardment [93]. Chemical vapor deposition is a thermodynamic process whose chemical reactions are under specific conditions of temperature, pressure, reaction rates, and momentum, mass, and energy transport [93, 94]. It involves the deposition of solid material from vapor by a chemical reaction occurring within the vicinity of the heated substrate. The solid material that results is in the form of films, crystal, or powder. This method allows one to tune a wide range of physical, chemical, and tribological properties of the material synthesized by varying conditions such as substrate temperature, substrate material, total pressure gas flow as well of the composition of the reaction gas mixture [94, 95]. All chemical vapor deposition reactions involve several steps which include the generation, transport, adsorption, decomposition of the precursor, migration of atoms to a strong binding site; nucleation leading to the growth of the thin film; and desorption and removal of unwanted side products [79, 96, 97].

4.2.1.2 Characterization

For nanocomposites whose water filtration applications involve adsorption, the surface morphology (microstructure) and particle size is an important property to consider since they affect adsorption properties. X-ray diffraction spectroscopy usually gives information on the crystalline structure of the material synthesized as well as the determination of the particle size [79], presenting an appropriate technique for the study of size modulation of the nanoparticles as a function of changes in parameters such as dopants, temperature, and synthesize time among others [68, 79]. Further, changes in the crystallinity of synthesized nanocomposites with varying the degree of doping can be studied using Raman spectroscopy [79] from the variations in the vibration modes of the nanomaterial due to the structural distortions caused by various dopant compositions [79, 98]. Transmission electron microscopy (TEM) and scanning electron microscopy (SEM) can also give micro-structural elucidation with varying dopant percentages as well as giving an estimate of the particle sizes [61, 79]. Furthermore, both TEM and SEM can give information on the level of aggregation or non-aggregation of the synthesized nanocomposites [79], as shown in figure 4.5.

X-ray photoelectron spectroscopy (XPS) is another technique used in the characterization of nanocomposites to give the atomic composition of the synthesized materials. This is important, especially because the properties of nanoparticles can greatly be tuned by changing the atomic composition of the core material or the dopant. The use of XPS, therefore, enables one to optimize the respective rations of each species in the nanocomposite. XPS can also be used to confirm the actual adsorption of target species (pollutant) on the surface of the adsorbent. This

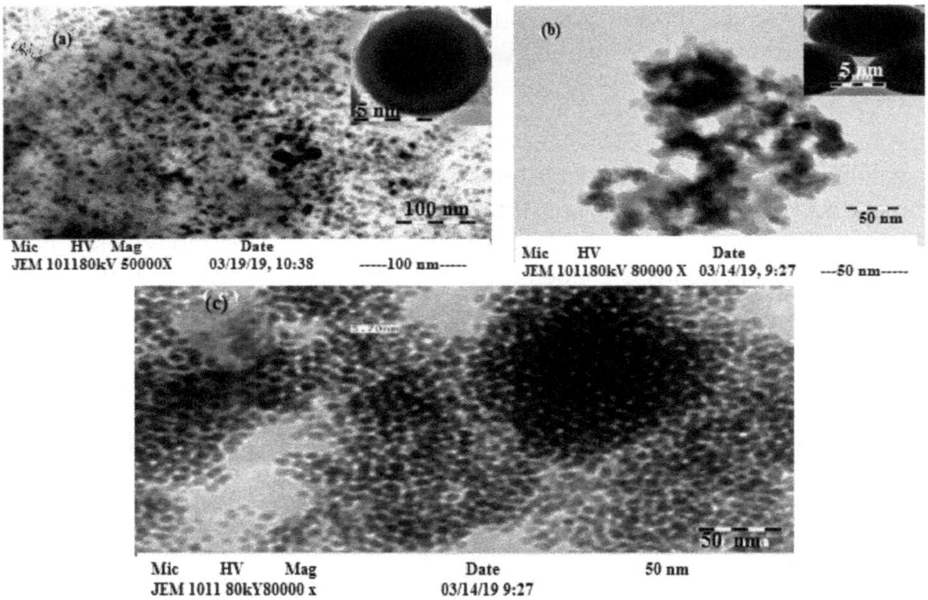

Figure 4.5. TEM micrographs of (a) Si-IL, (b) Fe_3O_4-Si-IL, and (c) Fe_3O_4. Reproduced from [99].

confirmation can also be complemented by the use of energy dispersive x-ray analysis (EDX). Further understanding of the inner architectural properties of nanocomposites is important since this determines the number of adsorption sites available. These can be inferred from the study of microporosity and specific surface areas using suitable adsorption isotherms such as the Langmuir adsorption isotherm, Freundlich adsorption models, or Brunauer–Deming–Deming–Teller [61, 68].

To probe the interaction between nanoparticles and bacteria, atomic force microscopy (AFM), TEM, and laser confocal microscopy (LCM) are used. This is confirmed by observing changes in the integrity of the cell membranes, resulting in the death of the bacteria. Inductively coupled plasma mass spectrometry (ICP-MS) may be used to quantify the concentration of silver ions leached from the nanocomposite matrix after a specified contact time with water samples containing the target microorganism [100].

Fourier transform infrared (FTIR) is also used to study the chemical interaction of various functional groups during synthesis and also to confirm the successful surface modification of the nanocomposites [66].

4.2.1.3 Application of non-polymeric nanocomposites in water purification

Magnesium oxide nanoparticles and magnesium nanoparticles exhibit effective biocidal activities against gram-positive bacteria such as *E. coli* and *Bacillus megaterium* as well as bacterial spores such as *Bacillus subtilis*. They are also known to absorb large amounts of halogen molecules from water. Similarly, silver nanoparticles are known antimicrobial compounds. This property is exhibited in the application of silver nanoparticles as antimicrobial agents against coliforms found in wastewater [92]. They possess biocidal activities against gram-positive and gram-negative bacteria such as *E. coli*, *Staphylococcus aureus*, *Klebsiella pneumoniae*, and *Pseudomonas aeruginosa* [53]. AuNPs have high adsorption capacity and can remove heavy toxic metals from water. For instance, in the removal of mercury from water, the Hg^{2+} are reduced to Hg metal, which forms an amalgam with AuNPs, which can easily be filtered out. The reduction of the Hg^{2+} may be designed in such a way that it takes place before coming into contact with the AuNPs or takes place on the surface of functionalized (stabilized) AuNPs. In the first case, reducing agents such as sodium borohydride ($NaBH_4$) are used to reduce Hg^{2+} to Hg metal before adsorption by the AuNPs. The Hg metal then adsorbs on the AuNP's surface forming Au_3Hg amalgam [101]. In the second design, the AuNPs are coated with a suitable reducing agent such as sodium citrate. In this method, the $Hg^{2+}$is reduced by citrate ions on the surface of AuNPs, followed by adsorption by the AuNPs to form Au_3Hg amalgam [102]. A schematic illustration of this method using sodium citrate as a stabilizer for AuNPs and a reducing agent for Hg^{2+} is shown in figure 4.6.

It is important to note that the above properties are exhibited by AuNPs but bulk gold does not possess these properties. Some metals such as tin are capable of removing mercury from water using the amalgam mechanism, both in bulk and nanoparticle form [103]. The choice of the substrate onto which the nanoparticles are immobilized affects its adsorption efficiency. For instance, immobilizing AuNPs on aluminum oxide (Al_2O_3) has been found to lower the adsorption of Hg^{2+} on citrate stabilized AuNPs [104] and unstabilized AuNPs [105]. However, AuNPs immobilized on Al_2O_3 show good adsorption for organomercury species such as methylmercury, ethylmercury, and phenylmercury [104]. Interestingly, the use of Al_2O_3 alone adsorbs poorly to organomercury species but adsorbs well to Hg^{2+}. Further, if the Al_2O_3 is immobilized into AuNPs, both organomercury species Hg^{2+} adsorb well [104].

Other examples of water filtration using metal and metal nanocomposites target the removal of dissolved organic compounds and organic liquids [101]. For example,

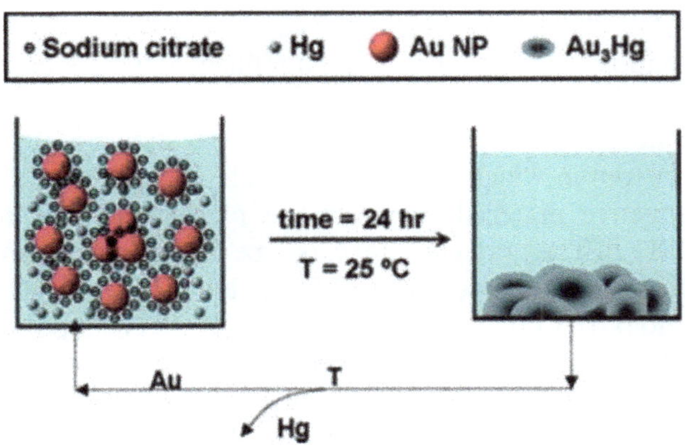

Figure 4.6. Removal of mercury from water using sodium citrate stabilized AuNPs. Adapted with permission from [102], Copyright 2012, American Chemical Society.

gold and silver nanoparticles form metal halide and amorphous carbon with halocarbons such as carbon tetrachloride (CCl_4) [106]. The end of the purification process in this case is signified by the change of color from that of AgNPs (blue) of AuNPs (red) to colorless, accompanied by the formation of an amorphous carbon solid, gray in color [106]. Other pollutants such as endosulfan, malathion, and chlorpyrifos pesticides can bind onto AuNPs; for instance, endosulfan and form aggregates, which settle at the bottom of the reaction vessel in form of a concentrate solid [107]. At room temperature, and using a water–methanol mixture solvent, chlorpyrifos is decomposed using AuNPs.

AgNPs also impregnate water filters for the purification of domestic water [80]. This application is usually accompanied by copper for the prevention of Legionella species colonization. Chlorine may also be used in conjunction with AgNPs especially in the sanitization of swimming pool waters [108]. The main advantage of this technique is that it does not produce unwanted oxidizing compounds, and has no odor or corrosion [80].

Magnetic $MnFe_2O_4$ nanoparticles coated with Mn–Co oxide shells exhibit good adsorption capacities toward heavy metals such as Pb^{2+}, Cd^{2+}, and Cu^{2+} [64]. The adsorption capacities of uncoated $MnFe_2O_4$ nanoparticles are found to decrease as the pH decreases. However, the Mn–Co oxide coated $MnFe_2O_4$ nanoparticles are excellent over a wide range of pH. Porous ZnO nanorods doped with aluminum are suitable adsorbents for dyes, hence presenting effective nanocomposites for the removal of dyes from wastewater [68]. In these nanocomposites, adsorption is found to increase with the increase in the amount of aluminum dopant. Undoped ZnO nanoparticles also possess high adsorption properties for heavy metals such as Zn^{2+}, Cd^{2+}, and Hg^{2+} and therefore can be used to remove these metals from wastewater [61]. Magnetite–silica nanocomposites can be used in water filtration to remove arsenic pollutants [66]. This is because magnetite nanoparticles bind arsenic (III) and arsenic (V) effectively [109]. The +3 and +5 oxidation states are the predominant forms of arsenic found in water [110].

4.2.2 Polymeric nanocomposites membranes

Polymer nanocomposites, also known as nanofilled polymers comprise a class of materials formed by at least a finely dispersed phase with nanomaterials such as nanoparticles, nanotubes, or nanosheets [111–113], as shown in figure 4.7. They are a superior type of tool in which the nanomaterials are spread in a polymer matrix resulting in innovative materials with unique physical and chemical properties [1, 111, 112, 114–119]. The polymer matrix can be made from a variety of polymers such as polyamides, polyethersulfone (PES), polyurethanes, polyamic acid, poly (vinylidene fluoride) (PVDF), poly(vinyl alcohol) (PVA), polyacrylonitrile, or polytetrafluoroethylene (PTFE) among others [111, 120]. The polymer offers special support for the nanomaterials as they usually possess tunable porous structures, excellent mechanical properties, and chemically bonded functional groups. These distinctive physical and chemical properties, lead to advanced morphologies and improved properties for the polymer nanocomposites such as high permselectivity,

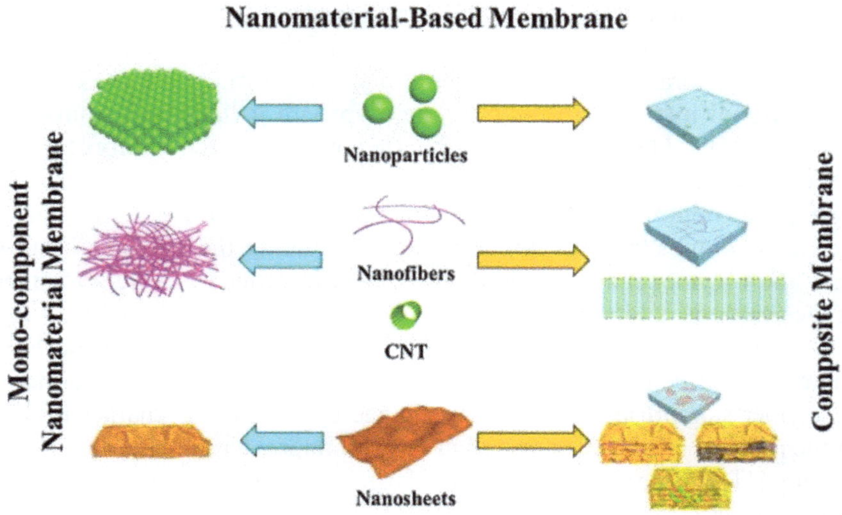

Figure 4.7. A schematic diagram representing the formation of nanomaterial-based membranes including their composites. Adapted from [113] figure 1. Copyright 2017, with permission from Elsevier.

fluxes, favorable surface morphology as a result of excellent fouling resistance compared with the pure polymeric matric membranes [121]. This usually enhances the performance of the composites due to the synergetic effect of the polymer and the nanomaterials giving it a high potential for various applications [111, 114, 116, 117]. In the recent past, polymer nanocomposite membranes have significantly enticed the attention of researchers for water treatment applications such as wastewater treatment, water purification, and removal of microorganisms, chemical compounds, and heavy metals among others. The integration of the different nanofillers, such as carbon nanotubes, zinc oxide, graphene oxide, silver and copper nanoparticles, titanium dioxide, and some other novel nanoscale materials into polymeric membranes provides great improvements such as increasing the hydrophilicity, subduing the accumulation of pollutants and foulants, enhancing rejection efficiencies and advancing mechanical properties and thermal stabilities of the polymer membrane [16]. The polymeric nanocomposite membranes are divided into thin-film nanocomposite membranes and blended nanocomposite membranes [122] normally prepared by mixing the nanosized fillers and the polymer matrix to enhance performances which can be done by physical blending or chemical polymerization technologies [16, 114, 121, 122]. In the next subsections, we will therefore discuss the synthesis, characterization, and application of these polymer nanocomposites membranes in water purification.

4.2.2.1 Synthesis of polymeric nanocomposite membranes

The development of high-performance polymer nanocomposites membranes usually requires thorough design, material selection, formulation procedure, and fabrication [118]. To achieve the required property, the selection of polymers and fillers/nanomaterials out of the vast number of polymers and fillers available and the determination of the best production process, considering the end-use requirements, and finally, fabrication of the material is very important [17, 118]. For example, one has to consider the compatibility of the polymer and the nanomaterial, the adhesion, and ease of incorporating the nanomaterial into the polymer, and the dispersion of the nanomaterial into the polymer to come up with a high performing polymer nanocomposite [17, 118]. In the polymer nanocomposites, the reinforcement phase requires uniform dispersion of the nanomaterial within the polymer matrix in order to maximize the reinforcing effect of the nanomaterials. Therefore in the synthesis of polymer nanocomposites, it is important to incorporate the individual nanomaterials into the polymer matrix and stabilize them in order to avoid agglomeration during the synthesis, and this requires the application of external energy of different forms to overcome the energy barrier within the nanomaterial bundles [118]. The nanomaterials can be placed within the membrane structure by either adding them to the polymer support layer in order to change the membrane characteristics or by adding them to the selective polymer skin layer to improve roughness, selectivity, and permeability [111]. Amongst the most popular methods of preparing polymeric nanocomposites are electro-spinning, track-etching, interfacial polymerization, stretching, and phase inversion. Figure 4.8 demonstrates the basic types of polymeric nanocomposite membranes. In the following sections, we discuss these common synthesis techniques and others with an emphasis on their advantages and disadvantages.

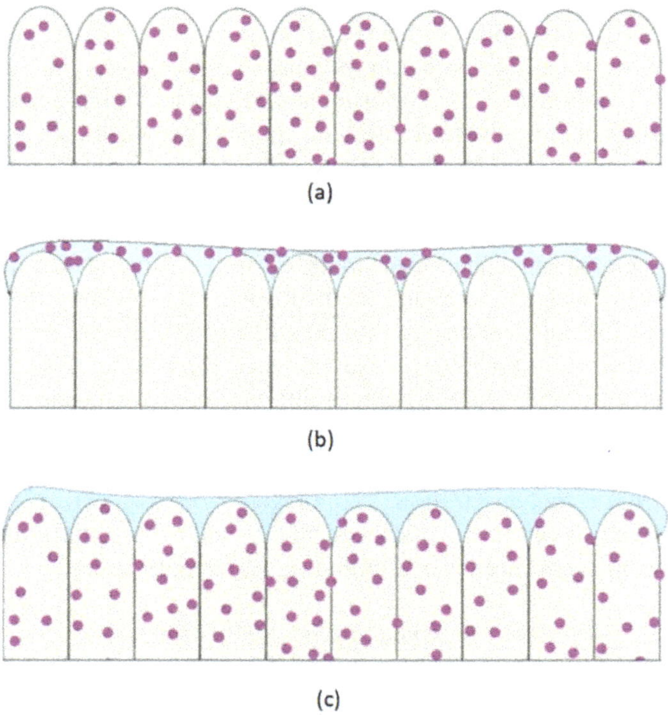

Figure 4.8. Types of nanocomposite membranes: (a) typical nanocomposite membrane; (b) nanocomposite membrane with nanoparticles in the skin layer; (c) nanocomposite membrane with nanoparticles in the support. Reproduced from [111], copyright 2019 Michał Adamczak *et al*, published by Hindawi. CC BY 4.0.

4.2.2.1.1 Phase inversion

Phase inversion, also known as phase separation is a process of controlled polymer transformation from a liquid phase to a solid phase [123–126]. It is the first breakthrough technique for membrane preparation to use a demixing approach that takes a polymer solution that is initially a homogenous liquid and uses control techniques to transform it into a solid [111, 123, 124, 126]. It has been mostly used to produce membranes for microfiltration, ultrafiltration, and nanofiltration [111]. Figure 4.9 shows a schematic diagram of phase inversion for nanocomposites.

There are four basic techniques which can be used to create phase inversion membranes, namely;

i. **Precipitation from the vapor phase.**
This process leads to a porous membrane from a solvent–polymer mixture which is cast on a film and placed in a vapor atmosphere containing a nonsolvent saturated with the same solvent. Membrane forms by diffusion of nonsolvent into the cast film due to the high concentration of solvent in the vapor atmosphere [123, 124].

ii. **Precipitation by controlled evaporation.**
In this method, a homogenous polymer solution is prepared from two or more volatile solvents of different boiling points and dissolution capacities [124] and then spread into a thin film [127]. Evaporation of the solvent from the cast polymer solution due to its high volatility occurs, inducing the polymer precipitation/demixing to form a skinned membrane [123, 124].

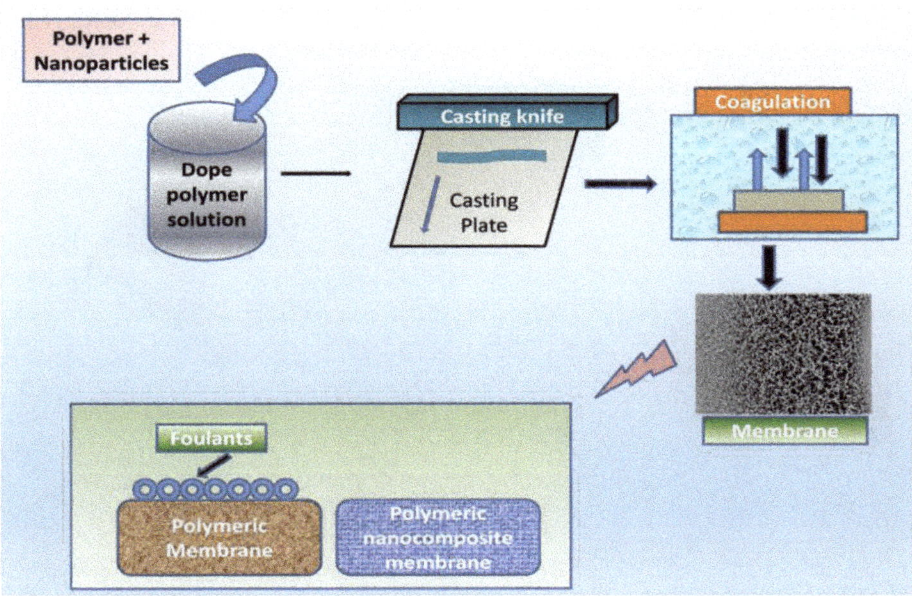

Figure 4.9. A schematic diagram representing the phase inversion technique. Reproduced from [122], copyright 2018 Zahid M *et al*, published by Longdom. CC BY.

iii. **Thermally induced phase separation.**
In this method, the polymer is dissolved in a suitable solvent at elevated temperatures to form a mixed or a single solvent polymer solution and then cast into a film [124]. The polymer solution is then cooled to a point at which precipitation of the polymer induces controlled thermal gelation. After complete precipitation, the solvent is extracted, evaporated, or freeze-dried [124]. The solvent evaporation induces membrane formation. This method is often used in preparing microfiltration membranes [123, 124].

iv. **Immersion precipitation.**
Phase inversion via immersion precipitation is the most widely-used membrane preparation method. A polymer plus solvent (polymer solution) is cast on a proper supporting layer and then submerged in a coagulation bath containing nonsolvent. Due to the solvent and nonsolvent exchange, precipitation takes place. The polymer must be soluble in the solvent mixture. The combination of phase separation and mass transfer affects the membrane structure [123, 124].

4.2.2.1.2 Electrospinning

Electrospinning, a relatively new technique is one of the most important techniques used to prepare polymeric nanocomposites [17, 117]. It has attracted a lot of interest since it is simple, inexpensive, and effective in producing polymeric nanocomposites with outstanding physicochemical properties such as high specific surface area, high porosity, and resistance against microorganisms [17, 117, 128]. The nanocomposite is produced by incorporating nanoparticles in a polymer solution which is then electrospun, resulting in nanoparticles-embedded polymeric fibers with enhanced or new properties [129]. The electrospinning equipment, as shown in figure 4.10, typically

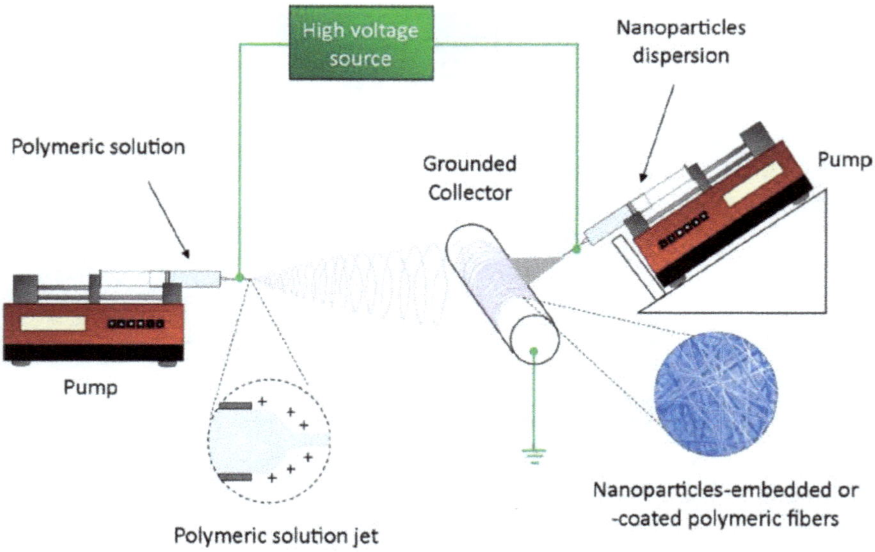

Figure 4.10. Typical configuration of electrospinning equipment. Reproduced from [129], copyright 2016 the authors, published by Intech. CC BY 3.0.

consists of a syringe containing a polymer solution passing through a capillary (needle) with the flow rate controlled by a pump and a power source which applies a high potential on the needle and therefore on the polymer solution [17, 129]. This applied voltage causes a significant density of charges on a droplet which protrudes into the tip of the needle until repulsive forces of the charges exceed the surface tension and a polymer jet or drop is eventually produced and then captured in the grounded collector, and the solvent is evaporated during the passage from the tip of the needle to the collector [129–131]. Studies have found that nanostructured morphology and fiber diameter can be significantly affected by applied potential strength, solution feed rates, the addition of ionic salt, and polymer solution viscosity, and depending on the molecular weight of the polymer selected, the minimum requirement for viscosity varies [17, 129, 131–136]. For example, the addition of salts such as sodium chloride, sodium/potassium phosphate has been shown to be useful in tuning the fiber diameter with the lowest average fiber diameter attained from poly (D,L-lactic acid) PDLA solution with 1 wt.% NaCl [17, 132–136].

4.2.2.1.3 Track etching

Track etching is an industrial process for the production of isoporous polymer membranes. It normally employs a nonporous polymeric film such as polycarbonate (PC) or polyethylene naphthalate film having a thickness in the 6 μm to 35 μm range which is exposed to radiation via accelerated heavy ions (charged particles) in a controlled manner [17, 137–139]. The charged particles pass across the film and leave behind a degraded (damaged) trajectory. This method is renowned for how it allows pore size distribution across the membrane to be precisely controlled [17]. Since the density and size of the pores are autonomous from each other, they may both be controlled across a broad spectrum from tens of micrometers down to just a few nanometers (size) and 1–1010 cm^{-2} (density) [17]. Besides, the density of the trajectories in the film depends on the time of exposure in the reactor and therefore a longer time in the reactor forms more tracks and causes larger pore density in the final membrane [124]. These properties normally simplify the relationship between the membrane structure and water transport properties [17]. Consequently, the membrane is treated with a caustic etchant such as sodium hydroxide solution to wash the damaged zone out from the trajectories to generate porosity [17, 124]. This way, the etching treatment widens the tracks into full-fledged pores, the size of which can be regulated precisely through the time of exposure, and the concentration and temperature of etchant solution [17, 124]. The membranes created with this technique are quite low in terms of porosity (up to 15%) and pore density; hence, the possibility of imperfections forming (double or triple pores) is reduced. Using these techniques, very precise pore size distribution is possible [139–141], as shown in figure 4.11.

4.2.2.1.4 Interfacial polymerization

Interfacial polymerization is a type of condensation polymerization where the polymerization occurs at an interface between an aqueous solution containing one monomer and an organic solution containing a second monomer [142]. It is the most

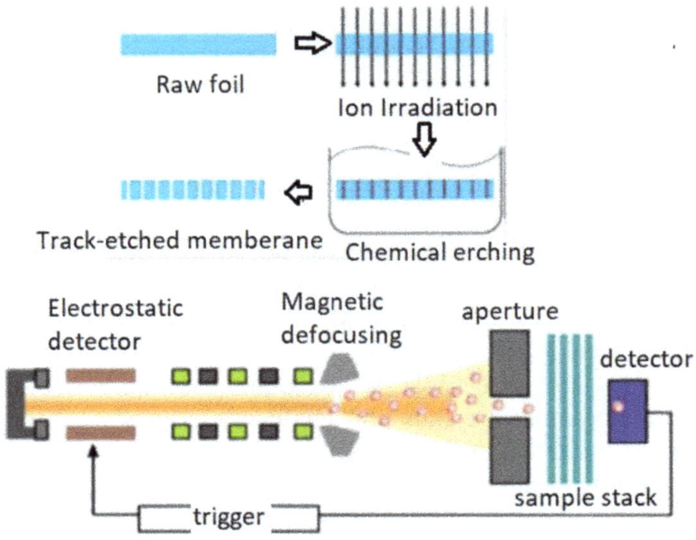

Figure 4.11. A scheme of single ion-irradiation used to fabricate track-etched membrane. Reproduced from [139], copyright 2018 Kamrani H *et al*, published by Longdom. CC BY.

significant technique for commercially fabricating thin-film composite, nanofiltration, and reverse osmosis [142] which uses two sets of reactive monomers reacted with each other in an organic/aqueous interface to create a capsule [124, 143].

4.2.2.1.5 In situ polymerization [144]

In situ polymerization, a solvent-free system is a chemical encapsulation technique that involves the mixing of nanomaterials in a neat monomer or a solution of monomer, followed by polymerization in the presence of the dispersed nanomaterials [145–148] leading to the covalent linkages between the polymer matrix and nanomaterial. It requires the utilization of monomers, an initiator, and a high-temperature reactor which empowers the achievement of homogeneous scattering of the nanomaterial in the polymer grid [149]. Figure 4.12 show a schematic diagram of *in situ* polymerization.

4.2.2.2 Characterization

Characterization of polymeric nanocomposite membranes is done in order to get information regarding the membrane performance after the modification of the membrane with nanomaterials [122]. The characterization is normally based on membrane performance (permeability, flux, rejection, and separation factors) and membrane morphology (pore size, pore-size distribution, the thickness of membrane, charge density, pore shape, chemical, and physical properties) which are key in the membrane performance [122]. This is done using common methods and techniques such as SEM, TEM, Fourier transform infrared spectroscopy (FTIR), x-ray diffraction (XRD), x-ray photo electron spectroscopy (XPS), thermo-gravimetric analysis (TGA), energy dispersive x-ray spectroscopy (EDX), dynamic light

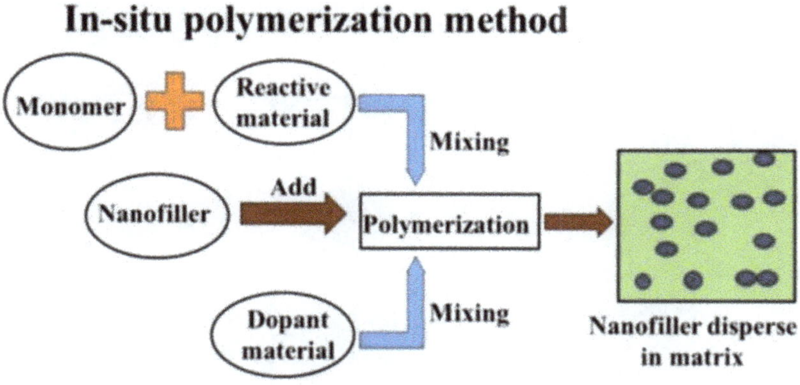

Figure 4.12. A schematic diagram showing the *In situ* polymerization method of preparation of conductive and insulating polymers. Reproduced from [150], copyright The Royal Society of Chemistry 2019. CC BY 3.0.

scattering (DLS), contact angle, porosity and zeta potential [4, 122]. In this section, we will briefly discuss some of these techniques.

4.2.2.2.1 Scanning electron microscopy (SEM)

SEM is used for the morphology and topography of the membrane surface [122]. It provides images of surface which are associated with sample properties such as homogeneity, roughness, porosity among others. The images are also used to get information about compatibility and lattice mismatch [4]. Usually, a narrow beam of electrons is scanned across the surface and deep inside of the membrane to produce an SEM image. This is due to the combination of inelastically scattered secondary electrons and elastically backscattered electrons. The deep penetration of the backscattered electrons provides valuable information regarding the composition of the membrane, while the secondary electrons provide information about the surface of the membrane [122, 151]. Figure 4.13 shows a cross-sectional SEM image of MWCNTs–polyamide nanocomposite membranes prepared with different nanotube concentrations demonstrating the negligible effect of MWCNT on the overall membrane thickness [152].

4.2.2.2.2 Transmission electron microscopy (TEM)

TEM is normally used to give qualitative and quantitative information about the internal structure, spatial distribution of the various phases, and views of the defective structure through direct visualization of the polymeric nanocomposite membranes [122, 153]. The principle of TEM involves transmitting electrons which give information regarding the size of nanoparticles [154–156] though it only characterizes thin film samples rather than the whole membrane [122].

4.2.2.2.3 Atomic force microscopy (AFM)

AFM is normally used to give the surface topology of the membrane by scanning over the sample using a cantilever with a sharp tip to get an image at atomic level [122, 157, 158]. Three modes, namely contact mode, non-contact mode, and tapping mode, are normally used to estimate the surface charges and Van der Waal forces

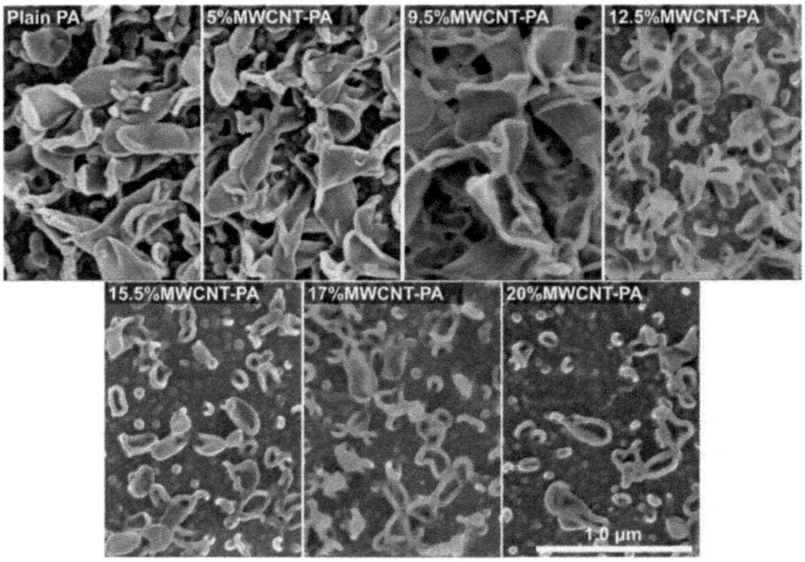

Figure 4.13. An unaltered SEM image of multi-walled carbon nanotubes (MWCNT)–polyamide (PA) nanocomposite membranes, for plain PA, and PA with 5, 9.5, 12.5, 15.5, 17, and 20 wt.% of MWCNT, where the typical lobe-like structures appear at the surface. Reproduced from [152], copyright the authors 2018, published by Springer Nature. CC BY 4.0.

between the tip and the membrane surface and eventually determine the roughness, pore size distribution, and phase on the surface membrane. AFM characterization can approximately provide information regarding properties such as fouling and antifouling of the membrane which depend on the surface smoothness [122].

4.2.2.2.4 Fourier transform infrared (FTIR)

FTIR technique is normally used to determine the cross-linking on the surface and to study the chemical structure of the membrane by analyzing the molecular vibrations following interaction between infrared (IR) radiations and the sample. It also provides information about the modified membrane whether it is functionalized or not and also gives information about the presence of new functional groups on the membrane surface [159, 160].

4.2.2.2.5 Thermal analytical methods

Thermal analytical techniques such as thermogravimetric analysis (TGA) have frequently been used to analyze the improvement of the processibility and thermal stability of the membrane due to the presence of polymer matrix [4, 161–165]. This method normally checks the thermal stability of the membranes by heating the membranes above room temperature ensuring that degradation, dissolution, and dissociation of the membrane do not occur.

4.2.2.2.6 X-ray photoelectron spectroscopy (XPS)

XPS, also known as elemental spectroscopy for chemical analysis (ESCA) is a qualitative and quantitative nondestructive method used to determine the chemical composition of the polymer membrane [122]. This is done by placing a dry sample under ultra-high vacuum and then focusing x-rays on the sample, which leads to an interaction of x-rays with sample causing photoemission of core electrons and binding energies of elements to be determined [166, 167]. It is said to be more sensitive to the membrane surface as compared to FTIR and is thus a powerful tool for the determination of the relative abundance of different functional groups in thin membrane layers, nanofiltration membrane structures, and modifications of membrane surfaces [167].

4.2.2.2.7 Other characterization methods

There are other methods which are used in the determination of other parameters such as pore radius, pore size, and porosity of the membrane. For example, the bubble point and liquid displacement porosimetry (LLDP) methods have been used to determine the pore radius and size distribution of the membrane. In the bubble point method, the original liquid is interchanged with air in the porous membrane to characterize the pore size and radius of the membrane while pressure is applied to replace the liquid with air. The porosity of the membrane can be characterized by the dry–wet method where the weight of dry and wet samples are measured and then a difference is taken of both dry and wet samples' weight [168, 169].

4.2.2.3 Applications of polymeric nanocomposite membranes in water purification

Water purification using polymeric membranes has shown a good capability for pollutant separation and hence it has been applied for water treatment. However, due to their limitations, their performance has been enhanced by the addition of nanofillers such as carbon nanotubes, zinc oxide, graphene oxide, AgNPs, and titanium dioxide into polymeric membranes have provided great advances, like enhanced hydrophilicity, suppression of accumulation of pollutants and foulants, enhanced rejection efficiencies and improved mechanical properties and thermal stabilities [16]. Consequently, different types of polymeric nanocomposite membranes are currently being explored for usage in water treatment applications such as wastewater treatment, water purification, removal of microorganisms, chemical compounds, and heavy metals among others [16], as illustrated in figure 4.14. The prime reason for the use of PNCs in any application lies in their unique properties, which are different from their counterparts [4]. In this subsection, we discuss various applications of polymeric nanocomposites in water purification.

4.2.2.3.1 Polymeric silver nanocomposite membranes

Materials based on AgNPs have been widely reported to offer antimicrobial properties that give them the potential for several applications such as water treatment since they are known to inhibit waterborne microorganisms such as algae, bacteria, and fungi [16, 17]. As a result, they have been employed in the fabrication of antibacterial membranes for water treatment purposes due to their thermal and chemical stability [17, 170]. For example, Ahmad *et al* [171] prepared a polymeric silver nanocomposite membrane via NIPs process using PES as the polymer matrix, with their results indicating an enhancement on the antibacterial and antifouling effect which they

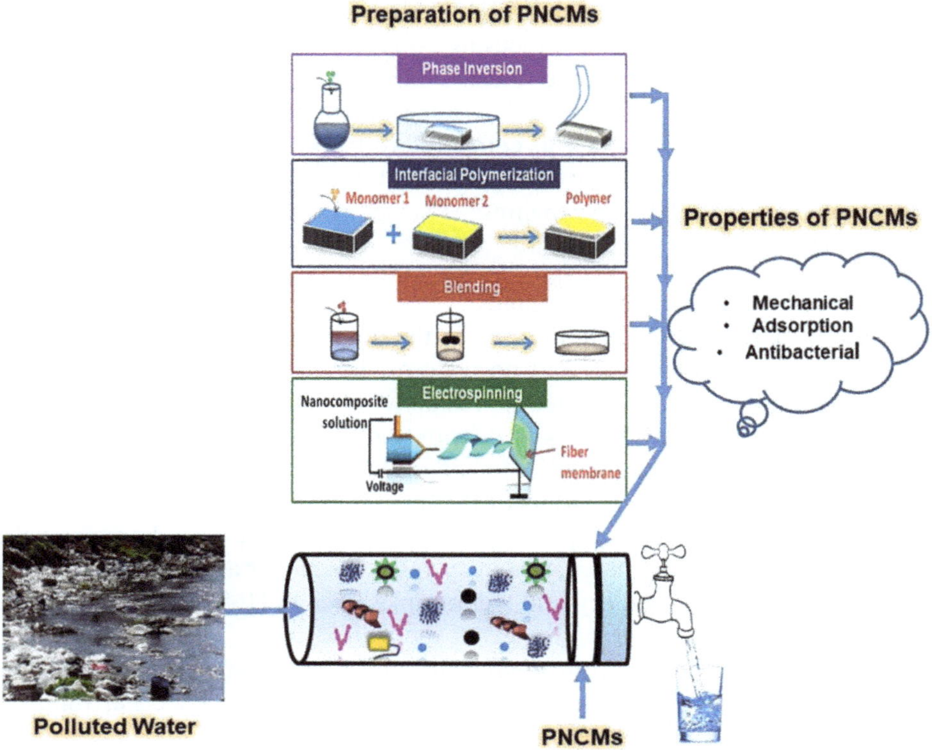

Figure 4.14. A schematic representation of membrane water treatment technology. Adapted from the graphical abstract of [19]. Copyright 2019, with permission from Elsevier.

attributed to the presence of the AgNPs on the surface of the membrane. Upon testing the nanocomposite membrane filtration of both bacteria-rich urban wastewater and bacteria poor seawater, they observed a reduced flux decline from 71% to 64% for the seawater and from 74% to 60.32% for the urban wastewater which again they attributed to the presence of the AgNPs in the PES matrix [171]. This is a good indication that this silver-based nanocomposite membrane can be fabricated and used as a water filtration membrane [171].

Tang *et al* [172], reported nanocomposite poly (vinylidene fluoride) (PVDF) membranes prepared through *in situ* formation of AgNPs which confirmed an increased antibacterial activity of the polymeric nanocomposite [172]. Likewise, a mixed matrix of the porous polymeric membrane with anti-fouling properties and enhanced permeate flux which was prepared via the phase-inversion method from a quaternary system of polysulfone/*N*,*N*-dimethylformamide/polyvinylpyrrolidone (PVP)/nanosilver is reported by Alpatova *et al* [172, 173]. Another study by Koseoglu-Imer *et al* [174] reported a polymeric membrane prepared from AgNPs and polysulfone (AgNp-PS) which was tested for water permeability and filtration performances using a model protein and carbohydrate solution as well as biofouling resistance by using a real activated sludge. Results from their study indicated that the addition of the AgNPs improved the protein and carbohydrate filtration performances of the AgNP-nanocomposite membrane as well as reduced absorptive and pore fouling values as compared to the bare PS membrane. The results further showed a minimum ionic silver loss from the composite membrane during pure water filtration which was measured using inductive-coupled plasma (ICP) spectrometer, an indication of a good filtration membrane [174]. In yet another study, mixed-matrix ultrafiltration (UF) membranes prepared by phase inversion of inorganic fillers such silver, copper, silica, zeolite, and silver–zeolite into a polysulfone (PS) as the polymeric dispersing matrix is reported by Hoek *et al* [175]. Their results indicated a simultaneous improvement in water permeability and solute selectivity of the membranes. The zeolite and silver-containing membranes were found to exhibit better protein fouling resistance (due to higher hydrophilicity), whereas silver and silver–zeolite based membranes produce better bacterial fouling resistance due to antimicrobial properties [175]. Their overall conclusion was that the zeolite-PSF and silver exchanged zeolite-PSF membranes offered the best combination of improved permeability, selectivity, and fouling resistance which was even superior to the commercial polyacrylonitrile (PAN) membranes [175].

Polymeric silver nanocomposites have been shown to enhance the biofouling of commercial nanofiltration (NF) membranes, as reported by Zhang *et al* [176]. In their study, they coated poly (vinyl alcohol) (PVA) onto the NF membrane and then deposited silver salt on the surface of the PVA layer. Their results indicated that, as compared to the pristine NF90 membrane, the PVA composite membranes (NF90-PVA) and AgNPs (NF90-PVA-AgNPs) modified membranes exhibited lower water flux, slightly higher salt rejection, strong antibacterial activity indicating that the NF90-PVA-AgNPs composite membrane displayed potential applications in industrial water-treatment [176]. Likewise, a thin-film polyamide reverses osmosis (RO) membrane modified using polydopamine (PDA) to uniformly disperse AgNPs inside the polymer layer forming a PDA-AgNP composite membrane has been reported by Yang *et al* [177]. The nanocomposite membrane was found to significantly enhance the antimicrobial effects of the silver loaded membranes on model bacteria *B. subtilis* and *E. coli* as well as enhanced salt rejection compared to the control PDA coated membrane. The authors attributed this to the preferential formation of AgNPs at defect sides within the PDA layer, a self-healing mechanism that can be used to prepare antimicrobial RO membranes with improved salt rejection without scarifying the membrane permeability, hence providing a new dimension for membrane surface modification [177]. Similarly, strong and stable polyamide nanofiltration membrane modified using biogenic AgNPs (BioAg0–6) as reported by Liu *et al* [178] were found to have increased hydrophilicity and water permeability, as high salt rejection [178]. The nanocomposite membrane also indicated a more excellent and longer-lasting antibacterial property which means that BioAg0–6 grafted TFC membranes could be potential as an effective strategy to decrease biofouling in the nanofiltration process [178].

4.2.2.3.2 Polymeric carbon nanotube (CNTs) nanocomposite membranes

Carbon nanotubes (CNTs) are said to have amazing electrical, mechanical, and thermal properties as well as partial antibacterial activity [16]. Due to these remarkable properties, they can change the physicochemical properties of membranes, which increase their potential for various applications [16]. Their inner pores usually tend to act as selective nanopores giving CNT-filled membranes enhanced permeability without a decrease in selectivity, as well as enhanced mechanical and thermal properties [179]. Because of these properties, CNTs have been used to make polymeric nanocomposites for water filtration. For example, as reported by Celik *et al* [180] a blended membrane of MWCNTs and PES synthesized via the phase inversion method was found to display a higher pure water flux and slower fouling rate as compared to pure PES membranes [180]. Similar results of enhanced hydrophilicity, improved membrane fouling resistance, and high pure water flux were also reported by Daraei *et al* [181] after modifying PES membrane by incorporating acid functionalized MWCNT. They concluded that the hyperbranched polycitric acid on MWCNTs presented numerous functional groups which significantly enriched membrane fouling resistance against whey proteins [181]. Furthermore, the increased hydrophilicity has also been found to increase rejection of hydrophobic pollutants as reported by Kim *et al* [182]. In their study, they fabricated polysulfone (PS) and polyamide (PA) thin-film nanocomposite membranes using chemically modified MWCNTs via phase inversion and interfacial polymerization methods, respectively. After evaluating the membrane systems for the removal of extractable organic fractions from sand process affected water (OSPW), their results displayed improved hydrophilicity which increased the rejection of hydrophobic pollutants, further increasing OSPW permeate flux, and significantly reducing membrane fouling [182].

Polymeric CNTs-composite membranes have been reported to remove heavy metals such as chromium and cadmium. For example, in a study by Shah and Murthy [183], multi-walled carbon nanotube/polysulfone (MWCNT/PS) composite membranes synthesized via the phase inversion appeared to be more hydrophilic, with a reduced flux, and lower flow rate when compared with pure polysulfone (PS) membranes. When the membranes were tested for heavy metal removal, they displayed maximum removals of about 94.2% and 78.2% for Cr(VI) and Cd(II), respectively, which was just 10.2% and 9.9%, respectively, when tested using unblended PS membranes [183]. Similarly, thin-film nanocomposite membranes prepared by embedding poly(methyl methacrylate) (PMMA) with hydrophobic modified multi-walled CNTs via interfacial polymerization presented a 99% Na_2SO_4 rejection with water permeate flux enhanced to about 62% compared to the pristine thin-film composite membrane, as reported by Shen *et al* [184]. Polyimide MWCNT composite membranes prepared by nonsolvent induced separation (NIPS) have also been found to enhance the performance of the filtration of dye solutions in water or ethanol concerning reference polymeric membranes as reported by Grosso *et al* [185].

Polymeric CNTs-composite membranes have been reported to improve the antibacterial activity making them very good candidates for drinking water filtration membranes. For example, coated nitrocellulose membranes with a nanocomposite containing 97% (wt.%) of polyvinyl-*N*-carbazole (PVK) and 3% (wt.%) of single-walled carbon nanotubes (SWNTs) (97:3 wt.% ratio PVK:SWNT) have been reported by Ahmed *et al* [186] to exhibit significant antimicrobial activity toward

Gram-positive and Gram-negative bacteria (~80%–90%); and a virus removal efficiency of ~2.5 logs. Additionally, the membranes were found to be nontoxic when tested against fibroblast cells as opposed to pure SWNT (100%) suggesting that they can be utilized for drinking water treatment.

Polymeric CNTs nanocomposite membranes have also been found to improve the anti-fouling properties of membrane bioreactors (MBR). As reported by Khalid *et al* [187] a better antifouling of polysulfone (PS) membrane was realized after blending with polyethylene glycol functionalized carbon nanotubes (PEG-CNTS) via non-solvent induced phase separation (NIPS) method. The PEG-CNTs composite PS when tested for wastewater treatment was found to reduce the interaction between protein and membrane surface thereby enhancing the fouling resistance by 79%. Similar results were reported by Mulopo *et al* [188] after treating bleach plant effluent in an anaerobic membrane reactor (AMBR) using CNTs–PSF membranes where they attributed the increased flux in the PS membranes to the presence of the O–H bonds which further showed the versatility of the CNTs-filled nanocomposite membranes for water treatment applications [188].

4.2.2.3.3 Polymeric titanium dioxide (TiO_2) nanocomposite membranes

Titanium dioxide (TiO_2) is a photocatalytic nanomaterial which has shown good thermal and chemical stability, and low human toxicity [164, 189] and as such has been widely studied for water disinfection and anti-biofouling. It is known to have an endless lifetime and generally remains unaffected during the degradation process of organic compounds and microorganisms [16]. In addition to its photo-induced hydrophilicity, it has high oxidation power which makes it potentially attractive for applications in membrane processes [16]. For example, a study by Teow *et al* [190] demonstrated anti-fouling properties of TiO_2NPs-filled polyvinylidene fluoride (PVDF) membranes prepared via *in situ* colloidal precipitation method, when tested for the filtration of humic acid (HA) as foulant in the presence of calcium ions (Ca^{2+}). Similarly, Nor *et al* [191] prepared a novel flat sheet PVDF nanocomposite membrane with TiO_2 nanofibers via the hot pressing method which showed excellent anti-fouling properties and high photocatalytic activity for the degradation of bisphenol A under UV irradiation. In yet another study by Zhang *et al* [192], they used a wet chemical method to synthesize TiO_2/PVDF composite membranes which displayed excellent permeability and anti-fouling performance indicating a great a potential for use in water purification applications as microfiltration membranes. Blended PES/TiO_2 flat-sheet membranes have also been found to enhance pure water flux, permeability, fouling resistance as well as solute rejection as reported by Sotto *et al* [193].

In addition to the anti-fouling properties, salt rejection by the polymeric TiO_2 nanocomposite membranes has also been reported. In a study by Mollahosseini and Rahimpour [194], they fabricated polysulfone (PS)/TiO_2 nanocomposite membrane via interfacial polymerization which showed that NaCl rejection was increased from 70% for neat nanofiltration membrane to 84% for 0.5 wt.% TiO_2 modified polysulfone nanofiltration membrane further indicating enhanced anti-fouling and permeability behavior of the prepared membranes. Also, the antibacterial property of prepared membranes was improved as a result of the photocatalytic characteristic of TiO_2 nanoparticles [194]. Similar results of salt rejection are reported by Zhang *et al* [195] where a TiO_2/PES nanostructured membrane exhibited a NaCl rejection higher than 90% which was four times greater than the PES substrate membrane. Another study Zhang *et al* [196] showed improved hydrophilicity and porosity of PES/TiO_2 nanocomposite membrane prepared via electrospinning technique, indicating a water flux improvement of 19% compared when compared with controlled PES membrane. All these and more studies indicate that the polymeric TiO_2 nanocomposite membranes can be used in water desalination [197, 198].

4.2.2.3.4 Polymeric graphene oxide nanocomposite membranes

Graphene oxide is a hydrophilic carbon nanomaterial obtained in the oxidized form of graphene [199]. It has been found to improve the thermal and mechanical properties of polymeric membranes [200] since it presents hydrophilic functional groups such as $-NH_2$, $-OH$, and $-SO_3H$ which provide a variety of surface-modification reactions [201, 202] that can be useful to functionalize graphene oxide- and graphene-based materials [203]. Due to all these properties, graphene has seen increased attention in nanocomposite membranes preparation for application in water treatment [204] as well as the removal of pharmaceutical traces from water and wastewater [205]. In this regards a report by Chang *et al* [206] indicated a synergistic effect of graphene oxide and polyvinylpyrrolidone (PVP) on PVDF ultrafiltration membrane performance which showed improved hydrophilicity and anti-fouling performances of the PVDF ultrafiltration membrane which they attributed to the addition of graphene oxide and PVP. The authors further reported that this improvement is due to the formation of hydrogen bonds between PVP and graphene oxide [206]. A similar synergistic effect on PVDF ultrafiltration membrane PAN membranes by graphene oxide and oxidized MWCNT which led to a significant increase of the permeability and anti-fouling performance of PVDF ultrafiltration membrane has been reported by Zhang *et al* [207, 208]. In both studies, composite membranes displayed a higher flux compared to other NF membranes when tested for the treatment of wastewater. Likewise, blending of copper oxide and graphene oxide on PVDF has been proposed as a potential composite material for water treatment due to enhanced anti-fouling property in terms of irreversible fouling based on the enhanced hydrophilic profile of the composite membrane, antibacterial activity and high effective permeability [209].

In another study, a composite of graphene and TiO_2 microsphere on the surface of a polymer filtration membrane was reported to show a high permeate flux as well as anti-fouling of the membrane [210] due to the hierarchical membrane structure, high photodegradation activity, and no membrane fouling. Indicating a promising potential in the clean water production field [210] as reported in the removal of dye molecules (methyl orange and rhodamine B) from water [211]. Similarly, improved water flux, as well as anti-fouling properties of poly(m-phenylene isophthalamide) (PMIA) nanofiltration membrane by addition of graphene oxide, has been reported by Yang *et al* [212]. Their results further indicated a 90% dye retention by the nanocomposite.

Polymeric graphene oxide nanocomposites have also seen possible application in water desalination. For example, a study by Ganesh *et al* [213] reported up to 72% removal of Na_2SO_4 from a feed solution by graphene oxide PSF nanofiltration (NF) membranes as well as an improved water flux when compared with the original polymeric membrane [213]. Similar high salt rejections for NaCl: 60%;and $CaCl_2$: >95% by graphene oxide-filled poly(amide-imide)-polyethylenimine (PAI-PEI) membrane was reported by Goh *et al* [214], as well as high rejection for different salts in the order: Na_2SO_4 (98.2%) > $MgSO_4$ (96.5%) > NaCl (56.8%) > $MgCl_2$ by polypiperazine-amide (PPA) composite NF membrane with graphene oxide reported by Wang *et al* [215]. All these studies point out the potential of the polymeric graphene oxide nanocomposite for water softening applications.

In addition to salt rejection, polymeric graphene oxide nanocomposites have also been used in the removal of heavy metals in water. For example, diallyldimethylammonium chloride (PDADMAC) in graphene oxide functionalized in PSF membranes was found to remove specific compounds such as heavy metals such as 86% Cu^{2+} and 88% Cd^{2+}, and salts including NaCl, Na_2SO_4, K_2SO_4, and $MgCl_2$ [216]. Likewise, Zhang *et al* [217] cross-linked a graphene oxide with isophorone diisocyanate (IPDI) and then coated the PVDF microfiltration (MF) membrane which led to 96% removal for dyes and 40%–70% removal of heavy metal ions (Pb^{2+}, Cu^{2+}, Cd^{2+}, and Cr^{3+}) in comparison with graphene oxide-PVDF membrane.

4.2.2.3.5 Polymeric zinc oxide (ZnO) nanocomposite membranes

Zinc oxide (ZnO), a multifunctional inorganic nanoparticle, has been found to have physical and chemical properties such as catalytic, antibacterial, and bactericide activities [218]. Due to these properties, they have been incorporated in polymers to enhance their hydrophilic, mechanical, and chemical properties [219, 220]. For example, the addition of ZnO was found to generate higher hydrophilicity on PES nanofiltration membranes, resulting in higher permeability and 254% flux improvement as compared to PES membranes [219] as well demonstrating the anti-fouling effect on the membrane surface as well as inside the pores of the composite membrane [218]. Furthermore, using zinc oxide led to enhanced fouling resistance during the filtration of solutions containing humic acid (HA), a typical foulant in natural waters [221]. Also, Liang *et al* [222] reported a 100% initial water permeability recovery after PVDF nanocomposite membranes filled with ZnO NPs cleaning meaning that the irreversible fouling feature, generally encountered in membrane processes, has been avoided. Likewise, polyvinylpyrrolidone (PVP) and ZnO nanocomposite membranes also displayed anti-fouling properties during removal of methylene blue and HA in water purification, as reported by Bai *et al* [223].

ZnO has also been used to form hybrids which have been found to have superior properties as compared to the ZnO composites. For example, a hybrid of ZnO and graphene oxide-PSF composite membrane reported by Chung *et al* [224] was found to demonstrate more hydrophilicity compared to other graphene oxide-PSF membranes, while exhibiting excellent anti-fouling and antibacterial properties during the filtration of HA solutions [224].

The enhanced antifouling properties of polymeric membranes by the addition of ZnO have been reported as a way of extending the shelf life of the membranes [199]. Polymeric ZnO nanocomposites membranes have shown high removal efficiency of heavy metals and thus they can be used in the treatment of wastewaters; they also tend to show high removal efficiencies of heavy metals [225]. All the discussed studies have clearly demonstrated that integration of ZnO into polymeric membranes to fabricate nanocomposites membranes leads to better surface hydrophilicity translated to better performance in terms of permeability and hence those membranes can be used in water quality analysis [226].

4.3 Conclusions and future directions

The growing deficiency of clean water due to the degradation of water quality from the release of large amounts of pollutants into water bodies has become an urgent global issue. For this reason, the treatment of polluted water is essential in order to satisfy the growing demands for sufficient clean water for human and industrial use. Nanotechnology has demonstrated a huge potential of addressing these challenges by presenting the opportunity to develop a wide variety of materials for water purification by modulating nanomaterial properties such as mechanical strength, dispersibility, porosity, hydrophobicity, and hydrophilicity among others. Significant steps have been achieved in the synthesis, characterization of nanocomposites for water treatment. There now exist proven nanotechnologies for water treatment and more are being developed. For example, as discussed in this chapter, a number of researchers have investigated and reported nanocomposites synthesis from metals, metal oxides, polymer, and membranes and their sound performance in water and wastewater treatment. Specifically, the nanocomposite membranes have shown a great role in water purification and reuse for various sources of water such as drinking water, brackish water, seawater, and wastewater treatment. It is important to note that the full potential in the use of nanocomposites for water treatment is not fully exploited and therefore there is room for more research on this topic.

Also, it is important to note that the scalability of nanocomposite production and associated costs are still a challenge. Further, some methods of synthesis such as solvothermal and thermal decomposition encompass harsh conditions such as high temperatures and organic solvents which would constitute pollution by themselves. In this regard, it would be important to embrace green synthesis at ordinary temperatures so as to afford eco-friendly, cost-effective, and sustainable technologies for water treatment. Synthesis methods such as precipitation/coprecipitation involve a number of steps such as preparation of core material, followed by stabilization and purification. This increases the cost and time of synthesis. It is necessary to develop single-step synthesis procedures where all these processes can be achieved in one step. Lastly, the sustainability of the nanocomposites has not been investigated and it would be important to look into their sustainability in water treatment.

References

[1] Beyene H D and Ambaye T G 2019 Application of sustainable nanocomposites for water purification process *Sustainable Polymer Composites and Nanocomposites* ed Inamuddin *et al* (Cham: Springer International Publishing) pp 387–412

[2] Ibrahim K *et al* 2020 Fabrication of nanostructured polyamic acid membranes for antimicrobially enhanced water purification *Adv. Polym. Tech.* **2020** 7362789

[3] Wang Z *et al* 2018 Recent advances in nanoporous membranes for water purification *Nanomaterials* **8** 65

[4] Pandey N, Shukla S K and Singh N B 2017 Water purification by polymer nanocomposites: an overview *Nanocomposites* **3** 47–66

[5] Tlili I and Alkanhal T A 2019 Nanotechnology for water purification: electrospun nanofibrous membrane in water and wastewater treatment *J. Water Reuse Desalination* **9** 232–48

[6] WHO/UNICEF 2014 *Progress on Drinking Water and Sanitation; Joint Monitoring Programme Update 2014* World Health Organization (WHO)/United Nations International Children's Fund (UNICEF)

[7] Camargo J A and Alonso Á 2006 Ecological and toxicological effects of inorganic nitrogen pollution in aquatic ecosystems: a global assessment *Environ. Int.* **32** 831–49

[8] Schaider L A *et al* 2014 Pharmaceuticals, perfluorosurfactants, and other organic wastewater compounds in public drinking water wells in a shallow sand and gravel aquifer *Sci. Total Environ.* **468–69** 384–93

[9] genome.gov *Genetic Disorders, Genomics and Healthcare (Neglected diseases)* Available from http://genome.gov/27531964

[10] Saxena S and Saxena U 2016 Development of bimetal oxide doped multifunctional polymer nanocomposite for water treatment *Int. Nano Lett.* **6** 223–34

[11] Inyinbor Adejumoke A *et al* 2018 Water pollution: effects, prevention, and climatic impact *Water Challenges of an Urbanizing World* ed M Glavan (Rijeka: IntechOpen)

[12] Yang Y *et al* 2018 Rapid water permeation through carbon nanomembranes with sub-nanometer channels *ACS Nano* **12** 4695–701

[13] The Water Project *Health and Water. Improving Health in Africa* Available from https://thewaterproject.org/why-water/health

[14] Safe Drinking Water Foundation 2017 *Conventional Water Treatment: Coagulation and Filtration* Available from https://safewater.org/fact-sheets-1/2017/1/23/conventional-water-treatment

[15] Mansouri J, Harrisson S and Chen V 2010 Strategies for controlling biofouling in membrane filtration systems: challenges and opportunities *J. Mater. Chem.* **20** 4567–86

[16] Ursino C *et al* 2018 Progress of nanocomposite membranes for water treatment *Membranes* **8** 18

[17] Bassyouni M *et al* 2019 A review of polymeric nanocomposite membranes for water purification *J. Ind. Eng. Chem.* **73** 19–46

[18] Yin K *et al* 2017 Femtosecond laser induced robust periodic nanoripple structured mesh for highly efficient oil–water separation *Nanoscale* **9** 14229–35

[19] Nasir A *et al* 2019 Progress in polymeric nanocomposite membranes for wastewater treatment: preparation, properties and applications *J. Ind. Eng. Chem.* **79** 29–40

[20] Wu C *et al* 2008 Treatment of oily water by a poly(vinyl alcohol) ultrafiltration membrane *Desalination* **225** 312–21

[21] Madaeni S S and Ghaemi N 2007 Characterization of self-cleaning RO membranes coated with TiO_2 particles under UV irradiation *J. Membr. Sci.* **303** 221–33

[22] Li R, Zhang L and Wang P 2015 Rational design of nanomaterials for water treatment *Nanoscale* **7** 17167–94

[23] Schwarzenbach R P *et al* 2010 Global water pollution and human health *Annu. Rev. Environ. Resour.* **35** 109–36

[24] Tang C *et al* 2015 Biomimetic aquaporin membranes coming of age *Desalination* **368** 89–105

[25] Ma N *et al* 2013 Nanocomposite substrates for controlling internal concentration polarization in forward osmosis membranes *J. Membr. Sci.* **441** 54–62

[26] Lakshmi K V, Nagrath K and Jha A 2011 *Access to Safe Water: Approaches for Nanotechnology Benefits to Reach the bottom of the Pyramid* Development Alternatives Group, New Delhi

[27] Andreescu S *et al* 2009 JEM spotlight: applications of advanced nanomaterials for environmental monitoring *J. Environ. Monit.* **11** 27–40

[28] Yazgan I *et al* 2014 Biofunctionalized poly (amic) acid membranes for absolute disinfection of drinking water *J. Membr. Sci.* **472** 261–71

[29] Pastoriza-Santos I and Liz-Marzán L M 2002 Synthesis of silver nanoprisms in DMF *Nano Lett.* **2** 903–5

[30] Sau T K and Murphy C J 2004 Room temperature, high-yield synthesis of multiple shapes of gold nanoparticles in aqueous solution *J. Am. Chem. Soc.* **126** 8648–49

[31] Sperling R A *et al* 2008 Biological applications of gold nanoparticles *Chem. Soc. Rev.* **37** 1896–908

[32] Shull K R and Kellock A J 1995 Metal particle adsorption and diffusion in a model polymer/metal composite system *J. Polym. Sci. B: Polym. Phys.* **33** 1417–22

[33] Vrijenhoek E M, Hong S and Elimelech M 2001 Influence of membrane surface properties on initial rate of colloidal fouling of reverse osmosis and nanofiltration membranes *J. Membr. Sci.* **188** 115–28

[34] Yang H-L, Lin J C-T and Huang C 2009 Application of nanosilver surface modification to RO membrane and spacer for mitigating biofouling in seawater desalination *Water Res.* **43** 3777–86

[35] Panigrahi S *et al* 2007 Synthesis and size-selective catalysis by supported gold nanoparticles: study on heterogeneous and homogeneous catalytic process *J. Phys. Chem.* C **111** 4596–605

[36] Okello V A *et al* 2011 Environmental applications of poly(amic acid)-based nanomaterials *J. Environ. Monit.* **13** 1236–45

[37] Gehrke I, Geiser A and Somborn-Schulz A 2015 Innovations in nanotechnology for water treatment *Nanotechnol. Sci. Appl.* **8** 1–17

[38] Ishida H, Campbell S and Blackwell J 2000 General approach to nanocomposite preparation *Chem. Mater.* **12** 1260–67

[39] Singh N *et al* 2017 Ceramic matrix composites: processing techniques and recent advancements *J. Mater. Environ. Sci.* **8** 1654–60

[40] Omanović-Mikličanin E *et al* 2020 Nanocomposites: a brief review *Health Technol.* **10** 51–9

[41] Sachinjith K R and Swathi K R 2018 A review on types of nanocomposites and their applications *Int. J. Adv. Res. Ideas Innov. Technol.* **4** 235–36

[42] Sharma G *et al* 2019 Novel development of nanoparticles to bimetallic nanoparticles and their composites: a review *J. King Saud Univ. Sci.* **31** 257–69

[43] Balaji V *et al* 2017 Nanocomposites and their applications *Int. J. Eng. Res. Mech. Civil Eng.* **2** 81–5

[44] Silvestre J, Silvestre N and de Brito J 2015 An overview on the improvement of mechanical properties of ceramics nanocomposites *J. Nanomater.* **2015** 106494

[45] Ng L Y *et al* 2013 Polymeric membranes incorporated with metal/metal oxide nanoparticles: a comprehensive review *Desalination* **308** 15–33

[46] Nguyen T P *et al* 2009 Hybrid nanocomposites for optical applications *Solid State Sci.* **11** 1810–14

[47] Sonawane G H, Patil S P and Sonawane S H 2018 Nanocomposites and its applications *Applications of Nanomaterials* ed S Mohan Bhagyaraj *et al* (Cambridge: Woodhead Publishing) pp 1–22

[48] Rathod V T, Kumar J S and Jain A 2017 Polymer and ceramic nanocomposites for aerospace applications *Appl. Nanosci.* **7** 519–48

[49] Lu H *et al* 2016 An overview of nanomaterials for water and wastewater treatment *Adv. Mater. Sci. Eng.* **2016** 4964828

[50] Mustapha S *et al* 2020 Application of TiO_2 and ZnO nanoparticles immobilized on clay in wastewater treatment: a review *Appl. Water Sci.* **10** 49

[51] Khan S U M, Al-Shahry M and Ingler W B 2002 Efficient photochemical water splitting by a chemically modified n-TiO_2 *Science* **297** 2243–45

[52] Rosa L R, Rosa R D and da Veiga M A M S 2016 Colloidal silver and silver nanoparticles bioaccessibility in drinking water filters *J. Environ. Chem. Eng.* **4** 3451–58

[53] Chaturvedi S and Dave P N 2019 Water purification using nanotechnology an emerging opportunities *Chem. Methodol.* **3** 115–44

[54] Goud R K *et al* 2018 Effect of metal oxide nanoparticles on Godavari river water treatment *AIP Conf. Proc.* **1966** 020003

[55] Šulek F *et al* 2010 Surface functionalization of silica-coated magnetic nanoparticles for covalent attachment of cholesterol oxidase *J. Magn. Magn. Mater.* **322** 179–85

[56] Wu M J *et al* 2014 Photocatalysis of titanium dioxide for water disinfection: challenges and future perspectives *Int. J. Photochem.* **2014** 973484

[57] Shahid M *et al* 2015 Magnetised titanium dioxide (TiO_2) for water purification: preparation, characterisation and application *Desalin. Water Treat.* **54** 979–1002

[58] Simeonidis K *et al* 2019 Implementing nanoparticles for competitive drinking water purification *Environ. Chem. Lett.* **17** 705–19

[59] O'Carroll D *et al* 2013 Nanoscale zero valent iron and bimetallic particles for contaminated site remediation *Adv. Water Res.* **51** 104–22

[60] Goveas J J *et al* 2020 Electrochemical synthesis of ZnO-WO_3 nanocomposites and their photocatalytic activity *J. Appl. Electrochem.* **50** 501–11

[61] Sani H A *et al* 2017 Nanocomposite of ZnO with montmorillonite for removal of lead and copper ions from aqueous solutions *Process Saf. Environ. Prot.* **109** 97–105

[62] Hosseinpour-Mashkani S M, Maddahfar M and Sobhani-Nasab A 2016 Precipitation synthesis, characterization, morphological control, and photocatalyst application of $ZnWO_4$ nanoparticles *J. Electron. Mater.* **45** 3612–20

[63] Lang J *et al* 2016 Chemical precipitation synthesis and significant enhancement in photocatalytic activity of Ce-doped ZnO nanoparticles *Ceram. Int.* **42** 14175–81

[64] Ma Z *et al* 2013 Synthesis of $MnFe_2O_4$@Mn–Co oxide core–shell nanoparticles and their excellent performance for heavy metal removal *Dalton Trans.* **42** 14261–67

[65] Kahrizi P, Mohseni-Shahri F S and Moeinpour F 2018 Adsorptive removal of cadmium from aqueous solutions using $NiFe_2O_4$/hydroxyapatite/graphene quantum dots as a novel nano-adsorbent *J. Nanostructure Chem.* **8** 441–52

[66] Kokate M, Garadkar K and Gole A 2013 One pot synthesis of magnetite–silica nanocomposites: applications as tags, entrapment matrix and in water purification *J. Mater. Chem.* A **1** 2022–29

[67] Zhong H, Mirkovic T and Scholes G D 2011 5.06 – Nanocrystal synthesis *Comprehensive Nanoscience and Technology* ed D L Andrews, G D Scholes and G P Wiederrecht (Amsterdam: Academic) pp 153–201

[68] Chouchene B *et al* 2017 Porous Al-doped ZnO rods with selective adsorption properties *Appl. Surf. Sci.* **409** 102–10

[69] Wetterskog E *et al* 2014 Precise control over shape and size of iron oxide nanocrystals suitable for assembly into ordered particle arrays *Sci. Technol. Adv. Mater.* **15** 055010

[70] Zhou Z *et al* 2015 Anisotropic shaped iron oxide nanostructures: controlled synthesis and proton relaxation shortening effects *Chem. Mater.* **27** 3505–15

[71] Kovalenko M V *et al* 2007 Fatty acid salts as stabilizers in size- and shape-controlled nanocrystal synthesis: the case of inverse spinel iron oxide *J. Am. Chem. Soc.* **129** 6352–53

[72] Hufschmid R *et al* 2015 Synthesis of phase-pure and monodisperse iron oxide nanoparticles by thermal decomposition *Nanoscale* **7** 11142–54

[73] Sun S and Zeng H 2002 Size-controlled synthesis of magnetite nanoparticles *J. Am. Chem. Soc.* **124** 8204–5

[74] Unni M *et al* 2017 Thermal decomposition synthesis of iron oxide nanoparticles with diminished magnetic dead layer by controlled addition of oxygen *ACS Nano* **11** 2284–303

[75] Wei X *et al* 2020 Facile ball-milling synthesis of CuO/Biochar nanocomposites for efficient removal of reactive red 120 *ACS Omega* **5** 5748–55

[76] Shan D *et al* 2016 Preparation of ultrafine magnetic biochar and activated carbon for pharmaceutical adsorption and subsequent degradation by ball milling *J. Hazard. Mater.* **305** 156–63

[77] Gu X *et al* 2016 Ball-milling synthesis of ZnO@sulphur/carbon nanotubes and Ni $(OH)_2$@sulphur/carbon nanotubes composites for high-performance lithium-sulphur batteries *Electrochim. Acta* **196** 369–76

[78] Lyu H *et al* 2018 Effects of ball milling on the physicochemical and sorptive properties of biochar: experimental observations and governing mechanisms *Environ. Pollut.* **233** 54–63

[79] Noah N M 2020 Design and synthesis of nanostructured materials for sensor applications *J. Nanomater.* **2020** 8855321

[80] Rus A, Leordean V-D and Berce P 2017 Silver nanoparticles (AgNP) impregnated filters in drinking water disinfection *MATEC Web Conf.* **137** 1–6

[81] Ayinde W B *et al* 2020 Green synthesis of AgMgOnHaP nanoparticles supported on chitosan matrix: defluoridation and antibacterial effects in groundwater *J. Environ. Chem. Eng.* **8** 104026

[82] Krithiga N, Rajalakshmi A and Jayachitra A 2015 Green synthesis of silver nanoparticles using leaf extracts of clitoria ternatea and solanum nigrum and study of its antibacterial effect against common nosocomial pathogens *J. Nanosci.* **2015** 928204

[83] Alizadeh H, Salouti M and Shapouri R 2014 Bactericidal effect of silver nanoparticles on intramacrophage *Brucella abortus* 544 *Jundishapur J. Microbiol.* **7** e9039
[84] Durán N *et al* 2016 Silver nanoparticles: a new view on mechanistic aspects on antimicrobial activity *Nanomed. Nanotechnol. Biol. Med.* **12** 789–99
[85] Mohanty S *et al* 2012 An investigation on the antibacterial, cytotoxic, and antibiofilm efficacy of starch-stabilized silver nanoparticles *Nanomed. Nanotechnol. Biol. Med.* **8** 916–24
[86] Matsumura Y *et al* 2003 Mode of bactericidal action of silver zeolite and its comparison with that of silver nitrate *Appl. Environ. Microbiol.* **69** 4278
[87] Nangmenyi G *et al* 2009 Synthesis and characterization of silver-nanoparticle-impregnated fiberglass and utility in water disinfection *Nanotechnology* **20** 495705
[88] El-Aassar A H M *et al* 2013 Using silver nanoparticles coated on activated carbon granules in columns for microbiological pollutants water disinfection in Abu Rawash area, Great Cairo, Egypt *Aust. J. Basic Appl. Sci.* **7** 422–32
[89] Dankovich T A and Gray D G 2011 Bactericidal paper impregnated with silver nanoparticles for point-of-use water treatment *Environ. Sci. Technol.* **45** 1992–98
[90] Praveena S M *et al* 2016 Preparation and characterisation of silver nanoparticle coated on cellulose paper: evaluation of their potential as antibacterial water filter *J. Exp. Nanosci.* **11** 1307–19
[91] Kallman E N, Oyanedel-Craver V A and Smith J A 2011 Ceramic filters impregnated with silver nanoparticles for point-of-use water treatment in rural Guatemala *J. Environ. Eng.* **137** 407–15
[92] Jain P and Pradeep T 2005 Potential of silver nanoparticle-coated polyurethane foam as an antibacterial water filter *Biotechnol. Bioeng.* **90** 59–63
[93] Abegunde O O *et al* 2019 Overview of thin film deposition techniques *Mater. Sci.* **6** 174–99
[94] Pedersen H and Elliott S D 2014 Studying chemical vapor deposition processes with theoretical chemistry *Theor. Chem. Acc.* **133** 1476
[95] Carlsson J-O and Martin P M 2010 Chemical vapor deposition *Handbook of Deposition Technologies for Films and Coatings* 3rd edn ed P M Martin (Boston, MA: William Andrew Publishing) pp 314–63
[96] Binions R and Parkin I P 2011 *Novel Chemical Vapour Deposition Routes to Nanocomposite Thin Films* (Rijeka: InTech Open)
[97] Ogawa F, Masuda C and Fujii H 2018 *In situ* chemical vapor deposition of metals on vapor-grown carbon fibers and fabrication of aluminum-matrix composites reinforced by coated fibers *J. Mater. Sci.* **53** 5036–50
[98] Vinodkumar R *et al* 2010 Highly conductive and transparent laser ablated nanostructured Al: ZnO thin films *Appl. Surf. Sci.* **257** 708–16
[99] Atta A M *et al* 2019 Novel magnetic silica-ionic liquid nanocomposites for wastewater treatment *Nanomaterials* 10 71
[100] Koslowski L A D *et al* 2018 Silver nanoparticles impregnated with polyamide-66 to disinfect drinking water *Rev. Ambient. Água* **13** 6
[101] Qian H *et al* 2013 Gold nanoparticles for cleaning contaminated water *J. Chem. Technol. Biotechnol.* **88** 735–41
[102] Ojea-Jiménez I *et al* 2012 Citrate-coated gold nanoparticles as smart scavengers for mercury (II) removal from polluted waters *ACS Nano* **6** 2253–60

[103] Biester H, Schuhmacher P and Müller G 2000 Effectiveness of mossy tin filters to remove mercury from aqueous solution by Hg(II) reduction and Hg(0) amalgamation *Water Res.* **34** 2031–36

[104] Lo S-I *et al* 2012 Gold nanoparticle–aluminum oxide adsorbent for efficient removal of mercury species from natural waters *Environ. Sci. Technol.* **46** 2724–30

[105] Lisha K P, Anshup and Pradeep T 2009 Towards a practical solution for removing inorganic mercury from drinking water using gold nanoparticles *Gold Bull.* **42** 144–52

[106] Nair A S and Pradeep T 2003 Halocarbon mineralization and catalytic destruction by metal nanoparticles *Curr. Sci. (Bangalore)* **84** 1560–63

[107] Bootharaju M S and Pradeep T 2012 Understanding the degradation pathway of the pesticide, chlorpyrifos by noble metal nanoparticles *Langmuir* **28** 2671–79

[108] Fewtrell L 2014 *Silver: Water Disinfection and Toxicity* Centre for Research into Environment and Health

[109] Tuutijärvi T *et al* 2009 As(V) adsorption on maghemite nanoparticles *J. Hazard. Mater.* **166** 1415–20

[110] Mayo J T *et al* 2007 The effect of nanocrystalline magnetite size on arsenic removal *Sci. Technol. Adv. Mater.* **8** 71–75

[111] Adamczak M, Kamińska G and Bohdziewicz J 2019 Preparation of polymer membranes by *in situ* interfacial polymerization *Int. J. Polym. Sci.* **2019** 6217924

[112] Passador F R, Ruvolo-Filho A and Pessan L A 2017 Nanocomposites of polymer matrices and lamellar clays *Nanostructures* ed A L Da Róz *et al* (Norwich, NY: William Andrew Publishing) pp 187–207

[113] Ying Y *et al* 2017 Recent advances of nanomaterial-based membrane for water purification *Appl. Mater. Today* **7** 144–58

[114] Huang T C, Yeh J M and Lai C Y 2012 Polymer nanocomposite coatings *Advances in Polymer Nanocomposites* ed F Gao (Cambridge: Woodhead Publishing) 605–38

[115] Tian T *et al* 2014 Graphene-based nanocomposite as an effective, multifunctional, and recyclable antibacterial agent *ACS Appl. Mater. Interfaces* **6** 8542–48

[116] Zhao S *et al* 2012 Performance improvement of polysulfone ultrafiltration membrane using well-dispersed polyaniline–poly(vinylpyrrolidone) nanocomposite as the additive *Ind. Eng. Chem. Res.* **51** 4661–72

[117] Yeum J H *et al* 2012 Carbon nanotube embedded multi-functional polymer nanocomposites, nanocomposites *Nanocomposites: New Trends and Developments* ed F Ebrahimi (Rijeka: IntetechOpen) pp 17–35

[118] Akpan E I *et al* 2019 Design and synthesis of polymer nanocomposites *Polymer Composites with Functionalized Nanoparticles* ed K Pielichowski and T M Majka (Amsterdam: Elsevier) pp 47–83

[119] Hanemann T and Szabó D V 2010 Polymer-nanoparticle composites: from synthesis to modern applications *Materials* **3** 3468–517

[120] Tang B, Huo Z and Wu P 2008 Study on a novel polyester composite nanofiltration membrane by interfacial polymerization of triethanolamine (TEOA) and trimesoyl chloride (TMC): I. Preparation, characterization and nanofiltration properties test of membrane *J. Membr. Sci.* **320** 198–205

[121] Singh M and Singh K 2017 Critical review on conducting and nanocomposite polymeric membranes for water purification *J. Basic Appl. Eng. Res.* **4** 292–95

[122] Zahid M *et al* 2018 A comprehensive review on polymeric nano-composite membranes for water treatment *J. Membrane Sci. Technol.* **8** 1–20

[123] Synder Filtration *Definition of Phase Inversion Membrane* Available from https://synderfiltration.com/learning-center/articles/introduction-to-membranes/phase-inversion-membranes-immersion-precipitation/

[124] Gohil J M and Choudhury R R 2019 Introduction to nanostructured and nano-enhanced polymeric membranes: preparation, function, and application for water purification *Nanoscale Materials in Water Purification* ed Thomas S *et al* (Amsterdam: Elsevier) pp 25–57

[125] Sadrzadeh M and Bhattacharjee S 2013 Rational design of phase inversion membranes by tailoring thermodynamics and kinetics of casting solution using polymer additives *J. Membr. Sci.* **441** 31–44

[126] Remanan S *et al* 2018 Recent advances in preparation of porous polymeric membranes by unique techniques and mitigation of fouling through surface modification *Chemistry Select* **3** 609–33

[127] Nguyen Q T *et al* 2010 Dry-cast process for synthetic microporous membranes: physico-chemical analyses through morphological studies *J. Membr. Sci.* **358** 13–25

[128] Liao Y *et al* 2018 Progress in electrospun polymeric nanofibrous membranes for water treatment: fabrication, modification and applications *Prog. Polym. Sci.* **77** 69–94

[129] Rodríguez-Tobías H, Morales G and Grande D 2016 Electrospinning and electrospraying techniques for designing antimicrobial polymeric biocomposite mats *Nanofiber Research – Reaching New Heights* ed M M Rahman and A M Asiri (Rijeka: IntechOpen)

[130] Lalia B S *et al* 2013 Fabrication and characterization of polyvinylidene fluoride-co-hexafluoropropylene (PVDF-HFP) electrospun membranes for direct contact membrane distillation *J. Membr. Sci.* **428** 104–15

[131] Gopal R *et al* 2006 Electrospun nanofibrous filtration membrane *J. Membr. Sci.* **281** 581–86

[132] Feng C *et al* 2008 Production of drinking water from saline water by air-gap membrane distillation using polyvinylidene fluoride nanofiber membrane *J. Membr. Sci.* **311** 1–6

[133] Zhou J *et al* 2018 Preparation and enhanced structural integrity of electrospun poly(ε-caprolactone)-based fibers by freezing amorphous chains through thiol-ene click reaction *Colloids Surf.* A **538** 7–13

[134] Wang C *et al* 2019 Facile preparation and catalytic performance characterization of AuNPs-loaded hierarchical electrospun composite fibers by solvent vapor annealing treatment *Colloids Surf.* A **561** 283–91

[135] Wang R *et al* 2012 Electrospun nanofibrous membranes for high flux microfiltration *J. Membr. Sci.* **392–93** 167–74

[136] Zhao Z *et al* 2012 High performance ultrafiltration membrane based on modified chitosan coating and electrospun nanofibrous PVDF scaffolds *J. Membr. Sci.* **394–95** 209–17

[137] Sartowska B *et al* 2013 Polymeric track etched membranes-application for advanced porous structures formation *Acta Phys. Pol.* A **123** 819–21

[138] Starosta W *et al* 1999 Investigations of heavy ion tracks in polyethylene naphthalate films *Radiat. Meas.* **31** 149–52

[139] Kamrani H and Nosrati A 2018 Fabrication of nanofiber filtration membranes using polyethylene terephthalate (PET): a review *J. Membrane Sci. Technol.* **8** 12

[140] Apel P 2001 Track etching technique in membrane technology *Radiat. Meas.* **34** 559–66

[141] Kim H G *et al* 2018 Propylene carbonate-derived size modulation of water cluster in pore-filled Nafion/polypropylene composite membrane for the use in vanadium redox flow batteries *J. Ind. Eng. Chem.* **60** 401–6

[142] Purkait M K *et al* 2018 Introduction to membranes *Interface Science and Technology* ed M K Purkait *et al* (Amsterdam: Elsevier) pp 1–37

[143] Perrier E and Hart J 2005 Smart vectorization: enzymatically activated encapsulation technologies *Delivery System Handbook for Personal Care and Cosmetic Products* ed M R Rosen (Norwich, NY: William Andrew Publishing) pp 797–816

[144] Madaeni S S, Zinadini S and Vatanpour V 2011 A new approach to improve antifouling property of PVDF membrane using *in situ* polymerization of PAA functionalized TiO_2 nanoparticles *J. Membr. Sci.* **380** 155–62

[145] Paszkiewicz S and Szymczyk A 2019 Graphene-based nanomaterials and their polymer nanocomposites *Nanomaterials and Polymer Nanocomposites* ed N Karak (Amsterdam: Elsevier) pp 177–216

[146] Nguyen T P and Yang S H 2018 Hybrid materials based on polymer nanocomposites for environmental applications *Polymer-based Nanocomposites for Energy and Environmental Applications* ed M Jawaid and M M Khan (Cambridge: Woodhead Publishing) pp 507–51

[147] Nelson G 2013 Microencapsulated colourants for technical textile application *Advances in the Dyeing and Finishing of Technical Textiles* ed M L Gulrajani (Cambridge: Woodhead Publishing) pp 78–104

[148] Pielichowski K and Michalowski S 2014 Nanostructured flame retardants: performance, toxicity, and environmental impact *Health and Environmental Safety of Nanomaterials* ed J Njuguna, K Pielichowski and H Zhu (Cambridge: Woodhead Publishing) pp 251–77

[149] Roy H *et al* 2019 Graphene-based nanovehicles for drug delivery *Characterization and Biology of Nanomaterials for Drug Delivery* ed S S Mohapatra *et al* (Amsterdam: Elsevier) pp 77–111

[150] Shukla V 2019 Review on the electromagnetic interference shielding materials fabricated by iron ingredients *Nanoscale Adv.* **1** 1640–71

[151] Rajeswari A, Vismaiya S and Pius A 2017 Preparation, characterization of nano ZnO-blended cellulose acetate-polyurethane membrane for photocatalytic degradation of dyes from water *Chem. Eng. J.* **313** 928–37

[152] Ortiz-Medina J *et al* 2018 Robust water desalination membranes against degradation using high loads of carbon nanotubes *Sci. Rep.* **8** 2748

[153] Bahmanyar M *et al* 2015 Preparation of ethylene vinyl acetate copolymer/graphene oxide nanocomposite films via solution casting method and determination of the mechanical properties *Polym. Plastics Technol. Eng.* **54** 218–222

[154] Jadav G L and Singh P S 2009 Synthesis of novel silica-polyamide nanocomposite membrane with enhanced properties *J. Membr. Sci.* **328** 257–67

[155] Ng L Y, Leo C P and Mohammad A W 2011 Optimizing the incorporation of silica nanoparticles in polysulfone/poly(vinyl alcohol) membranes with response surface methodology *J. Appl. Polym. Sci.* **121** 1804–14

[156] Mokhtari S *et al* 2017 Enhancing performance and surface antifouling properties of polysulfone ultrafiltration membranes with salicylate-alumoxane nanoparticles *Appl. Surf. Sci.* **393** 93–102

[157] Magonov S N and Reneker D H 1997 Characterization of polymer surfaces with atomic force microscopy *Annu. Rev. Mater. Sci.* **27** 175–222

[158] Wan Y *et al* 2015 Mechanical properties and cytotoxicity of nanoplate-like hydroxyapatite/polylactide nanocomposites prepared by intercalation technique *J. Mech. Behav. Biomed. Mater.* **47** 29–37
[159] Hong J and He Y 2014 Polyvinylidene fluoride ultrafiltration membrane blended with nano-ZnO particle for photo-catalysis self-cleaning *Desalination* **332** 67–75
[160] Al-Hinai M H *et al* 2017 Antimicrobial activity enhancement of poly(ether sulfone) membranes by *in situ* growth of ZnO nanorods *ACS Omega* **2** 3157–67
[161] Meneghetti P and Qutubuddin S 2006 Synthesis, thermal properties and applications of polymer-clay nanocomposites *Thermochim. Acta* **442** 74–77
[162] Zheng J *et al* 2017 Sulfonated multiwall carbon nanotubes assisted thin-film nanocomposite membrane with enhanced water flux and anti-fouling property *J. Membr. Sci.* **524** 344–53
[163] Yang E *et al* 2017 Enhanced desalination performance of forward osmosis membranes based on reduced graphene oxide laminates coated with hydrophilic polydopamine *Carbon* **117** 293–300
[164] Fan Y *et al* 2017 Poly(vinylidene fluoride) hollow-fiber membranes containing silver/graphene oxide dope with excellent filtration performance *J. Appl. Polym. Sci.* **134** 15
[165] Davood Abadi Farahani M H, Hua D and Chung T-S 2017 Cross-linked mixed matrix membranes consisting of carboxyl-functionalized multi-walled carbon nanotubes and P84 polyimide for organic solvent nanofiltration (OSN) *Sep. Purif. Technol.* **186** 243–54
[166] Lin Y *et al* 2017 Preparation of high-performance Al_2O_3/PES composite hollow fiber UF membranes via facile *in situ* vapor induced hydrolyzation *J. Membr. Sci.* **539** 65–75
[167] Yan L *et al* 2009 Application of the Al_2O_3–PVDF nanocomposite tubular ultrafiltration (UF) membrane for oily wastewater treatment and its antifouling research *Sep. Purif. Technol.* **66** 347–52
[168] Ihsanullah *et al* 2016 Fabrication and antifouling behaviour of a carbon nanotube membrane *Mater. Des.* **89** 549–58
[169] Wang H *et al* 2018 Enhancement of hydrophilicity and the resistance for irreversible fouling of polysulfone (PSF) membrane immobilized with graphene oxide (GO) through chloromethylated and quaternized reaction *Chem. Eng. J.* **334** 2068–78
[170] López-Heras M *et al* 2015 Towards understanding the antibacterial activity of Ag nanoparticles: electron microscopy in the analysis of the materials-biology interface in the lung *Environ. Sci.: Nano* **2** 312–26
[171] Ahmad Rehan Z *et al* 2016 Synthesis and characterization of silver nanoparticles-filled polyethersulfone membranes for antibacterial and anti-biofouling application *Recent Pat. Nanotechnol.* **10** 231–51
[172] Tang X and Cao X 2012 Preparation and characterization of antibacterial poly (vinylidene fluoride)-silver composites *High Perform. Polym.* **24** 135–39
[173] Alpatova A *et al* 2013 Fabrication of porous polymeric nanocomposite membranes with enhanced anti-fouling properties: effect of casting composition *J. Membr. Sci.* **444** 449–60
[174] Koseoglu-Imer D Y *et al* 2013 The production of polysulfone (PS) membrane with silver nanoparticles (AgNP): physical properties, filtration performances, and biofouling resistances of membranes *J. Membr. Sci.* **428** 620–28
[175] Hoek E M V *et al* 2011 Physical–chemical properties, separation performance, and fouling resistance of mixed-matrix ultrafiltration membranes *Desalination* **283** 89–99
[176] Zhang Y *et al* 2016 Facile modification of thin-film composite nanofiltration membrane with silver nanoparticles for anti-biofouling *J. Polym. Res.* **23** 105

[177] Yang Z *et al* 2016 *In situ* reduction of silver by polydopamine: a novel antimicrobial modification of a thin-film composite polyamide membrane *Environ. Sci. Technol.* **50** 9543–50

[178] Liu S *et al* 2015 The anti-biofouling properties of thin-film composite nanofiltration membranes grafted with biogenic silver nanoparticles *Desalination* **375** 121–28

[179] Fontananova E *et al* 2017 Effect of functional groups on the properties of multiwalled carbon nanotubes/polyvinylidenefluoride composite membranes *J. Membrane Sci.* **541** 198–204

[180] Celik E *et al* 2011 Carbon nanotube blended polyethersulfone membranes for fouling control in water treatment *Water Res.* **45** 274–82

[181] Daraei P *et al* 2013 Enhancing antifouling capability of PES membrane via mixing with various types of polymer modified multi-walled carbon nanotube *J. Membrane Sci.* **444** 184–91

[182] Kim E-S, Liu Y and Gamal El-Din M 2013 An *in situ* integrated system of carbon nanotubes nanocomposite membrane for oil sands process-affected water treatment *J. Membr. Sci.* **429** 418–27

[183] Shah P and Murthy C N 2013 Studies on the porosity control of MWCNT/polysulfone composite membrane and its effect on metal removal *J. Membr. Sci.* **437** 90–8

[184] Shen J *et al* 2013 Preparation and characterization of thin-film nanocomposite membranes embedded with poly (methyl methacrylate) hydrophobic modified multiwalled carbon nanotubes by interfacial polymerization *J. Membr. Sci.* **442** 18–26

[185] Grosso V *et al* 2014 Polymeric and mixed matrix polyimide membranes *Sep. Purif. Technol.* **132** 684–96

[186] Ahmed F *et al* 2013 Antimicrobial PVK: SWNT nanocomposite coated membrane for water purification: performance and toxicity testing *Water Res.* **47** 3966–75

[187] Khalid A *et al* 2018 PEG-CNTs nanocomposite PSU membranes for wastewater treatment by membrane bioreactor *Sep. Purif. Technol.* **190** 165–76

[188] Mulopo J 2017 Bleach plant effluent treatment in anaerobic membrane bioreactor (AMBR) using carbon nanotube/polysulfone nanocomposite membranes *J. Environ. Chem. Eng.* **5** 4381–87

[189] Marino T *et al* 2017 Photocatalytic activity and synthesis procedures of TiO_2 nanoparticles for potential applications in membranes *Application of Nanotechnology in Membranes for Water Treatment* (Boca Raton, FL: CRC Press)

[190] Teow Y *et al* 2013 Studies on the surface properties of mixed-matrix membrane and its antifouling properties for humic acid removal *J. Appl. Polym. Sci.* **128** 3184–92

[191] Nor N *et al* 2016 Preparation and performance of PVDF-based nanocomposite membrane consisting of TiO_2 nanofibers for organic pollutant decomposition in wastewater under UV irradiation *Desalination* **391** 89–97

[192] Zhang W *et al* 2016 A facile TiO_2/PVDF composite membrane synthesis and their application in water purification *J. Nanopart. Res.* **18** 31

[193] Sotto A *et al* 2011 Doping of polyethersulfone nanofiltration membranes: antifouling effect observed at ultralow concentrations of TiO_2 nanoparticles *J. Mater. Chem.* **21** 10311–20

[194] Mollahosseini A and Rahimpour A 2014 Interfacially polymerized thin film nanofiltration membranes on TiO_2 coated polysulfone substrate *J. Ind. Eng. Chem.* **20** 1261–68

[195] Alam J *et al* 2016 Atomic layer deposition of TiO_2 film on a polyethersulfone membrane: separation applications *J. Polym. Res.* **23** 183

[196] Zhang C *et al* 2017 Electrospun polysulfone (PSf)/titanium dioxide (TiO_2) nanocomposite fibers as substrates to prepare thin film forward osmosis membranes *J. Chem. Technol. Biotechnol.* **92** 2090–97

[197] Amini M, Rahimpour A and Jahanshahi M 2016 Forward osmosis application of modified TiO_2-polyamide thin film nanocomposite membranes *Desalin. Water Treat.* **57** 14013–23

[198] Emadzadeh D *et al* 2014 A novel thin film composite forward osmosis membrane prepared from PSf–TiO_2 nanocomposite substrate for water desalination *Chem. Eng. J.* **237** 70–80

[199] Jhaveri J H and Murthy Z V P 2016 Nanocomposite membranes *Desalin. Water Treat.* **57** 26803–19

[200] Ionita M *et al* 2014 Improving the thermal and mechanical properties of polysulfone by incorporation of graphene oxide *Compos. B: Eng.* **59** 133–39

[201] Enotiadis A *et al* 2012 Graphene-based Nafion nanocomposite membranes: enhanced proton transport and water retention by novel organo-functionalized graphene oxide nanosheets *Small* **8** 3338–49

[202] Liu G *et al* 2017 Graphene oxide/triethanolamine modified titanate nanowires as photocatalytic membrane for water treatment *Chem. Eng. J.* **320** 74–80

[203] Jhaveri J H and Murthy Z V P 2016 A comprehensive review on anti-fouling nanocomposite membranes for pressure driven membrane separation processes *Desalination* **379** 137–54

[204] An D *et al* 2016 Separation performance of graphene oxide membrane in aqueous solution *Ind. Eng. Chem. Res.* **55** 4803–10

[205] Carmalin Sophia A *et al* 2016 Application of graphene based materials for adsorption of pharmaceutical traces from water and wastewater – a review *Desalin. Water Treat.* **57** 27573–86

[206] Chang X *et al* 2014 Exploring the synergetic effects of graphene oxide (GO) and polyvinylpyrrodione (PVP) on poly(vinylylidenefluoride) (PVDF) ultrafiltration membrane performance *Appl. Surf. Sci.* **316** 537–48

[207] Zhang J *et al* 2013 Synergetic effects of oxidized carbon nanotubes and graphene oxide on fouling control and anti-fouling mechanism of polyvinylidene fluoride ultrafiltration membranes *J. Membr. Sci.* **448** 81–92

[208] Zhang L *et al* 2016 High flux MWCNTs-interlinked GO hybrid membranes survived in cross-flow filtration for the treatment of strontium-containing wastewater *J. Hazard. Mater.* **320** 187–93

[209] Zhao C *et al* 2017 Highly antifouling and antibacterial performance of poly (vinylidene fluoride) ultrafiltration membranes blending with copper oxide and graphene oxide nanofillers for effective wastewater treatment *J. Colloid Interface Sci.* **505** 341–51

[210] Gao P *et al* 2013 Multifunctional graphene oxide–TiO_2 microsphere hierarchical membrane for clean water production *Appl. Catal.* B **138–39** 17–25

[211] Xu C *et al* 2013 Graphene oxide–TiO_2 composite filtration membranes and their potential application for water purification *Carbon* **62** 465–71

[212] Yang M *et al* 2017 Preparation of graphene oxide modified poly(m-phenylene isophthalamide) nanofiltration membrane with improved water flux and antifouling property *Appl. Surf. Sci.* **394** 149–59

[213] Ganesh B M, Isloor A M and Ismail A F 2013 Enhanced hydrophilicity and salt rejection study of graphene oxide-polysulfone mixed matrix membrane *Desalination* **313** 199–207

[214] Goh K *et al* 2015 Graphene oxide as effective selective barriers on a hollow fiber membrane for water treatment process *J. Membr. Sci.* **474** 244–53

[215] Wang J *et al* 2016 Graphene oxide polypiperazine-amide nanofiltration membrane for improving flux and anti-fouling in water purification *RSC Adv.* **6** 82174–85

[216] Kochameshki M G *et al* 2017 Grafting of diallyldimethylammonium chloride on graphene oxide by RAFT polymerization for modification of nanocomposite polysulfone membranes using in water treatment *Chem. Eng. J.* **309** 206–21

[217] Zhang P *et al* 2017 Cross-linking to prepare composite graphene oxide-framework membranes with high-flux for dyes and heavy metal ions removal *Chem. Eng. J.* **322** 657–66

[218] Leo C P *et al* 2012 Polysulfone membranes blended with ZnO nanoparticles for reducing fouling by oleic acid *Sep. Purif. Technol.* **89** 51–6

[219] Shen L *et al* 2012 Preparation and characterization of ZnO/polyethersulfone (PES) hybrid membranes *Desalination* **293** 21–9

[220] Lin W *et al* 2009 Toxicity of nano- and micro-sized ZnO particles in human lung epithelial cells *J. Nanopart. Res.* **11** 25–39

[221] Balta S *et al* 2012 A new outlook on membrane enhancement with nanoparticles: the alternative of ZnO *J. Membr. Sci.* **389** 155–61

[222] Liang S *et al* 2012 A novel ZnO nanoparticle blended polyvinylidene fluoride membrane for anti-irreversible fouling *J. Membr. Sci.* **394–95** 184–92

[223] Bai H, Liu Z and Sun D D 2012 A hierarchically structured and multifunctional membrane for water treatment *Appl. Catal.* B **111–12** 571–77

[224] Chung Y T *et al* 2017 Development of polysulfone-nanohybrid membranes using ZnO-GO composite for enhanced antifouling and antibacterial control *Desalination* **402** 123–32

[225] Khan S B *et al* 2015 Assessment of antibacterial cellulose nanocomposites for water permeability and salt rejection *J. Ind. Eng. Chem.* **24** 266–75

[226] Wang Y *et al* 2009 Preparation of cellulose acetate membrane filled with metal oxide particles for the pervaporation separation of methanol/methyl tert-butyl ether mixtures *Chem. Eng. J.* **146** 6–10

Chapter 5

Advancing water quality assessment via artificial neural networks (ANNs)

Sachchidanand Soaham Gupta, Rajeev Singh and Pratibha Chaudhary

Water is very rightly dubbed the 'elixir of life'. Predicting water quality plays a significant role in environmental protection, preservation of the habitat, and aquaculture. Conventionally, mathematical and statistical models designed on the theories of analytical chemistry have been applied in analysis and monitoring of the quality of water resources. In recent years, the world has seen an exponential growth in artificial intelligence (AI) and machine learning (ML). Developments in artificial neural networks (ANNs) have made them an access point for the utilization of statistical models of ML in water quality analysis and monitoring. There are numerous water quality parameters that have been identified and categorized as chemical, physical, and biological variables. These variable data can be obtained using various physico-chemical sensors and techniques. In this chapter we study the various water quality parameters and the fundamentals of ML along with ANN models utilized in real-time for water quality assessment to create a vivid understanding of the topic. Various new architectures are also discussed such as hybrid, recurrent and feedforward which can enhance the modeling efficiency of future technologies.

5.1 Introduction

Water is the most valuable resource to humans. It has supported life as we see today on earth. The first forms of life had evolved from aquatic ecosystems and then moved to terrestrial forms through evolutionary processes and natural selection. One of the first human civilizations, the Indus Valley Civilization grew on the banks of the great river Indus and its tributaries. The Gangetic plains, situated in the north of India, are still one of the most densely populated places on earth owing to the fertile lands on the banks of the river Ganga and its tributaries. All these facts combine to signify a single fact—*'water is precious to human life'*.

doi:10.1088/978-0-7503-3047-3ch5

With the global population rising exponentially in the last century, the stress on limited natural resources has increased. Water amongst these is one of the most overstressed resources. According to a study conducted by the US Environment Protection Agency in 2009, it has been estimated that around 3.6 billion people globally already live in potential water scarcity at least once every month and this figure has been predicted to rise to around 4.8–5.7 billion by the year 2050 [1]. The World Economic Forum in its *Global Risk Report* for the year 2020 has listed water scarcity as the seventh most likely and fifth worst impacting risk [2].

The deterioration of water quality in water bodies such as rivers, ponds, and lakes etc, because of numerous environmental as well as anthropological factors, directly impacts the value of these water bodies. Precise water quality assessment and prediction hence plays a key role in environmental sustainability, human health, and ecosystem balance monitoring. Furthermore, prediction of potential changes in water quality in the near future means we can improve the measures taken for protection of these resources within time and control the damage done [3]. Many organizations and boards monitor water quality variables in order to guarantee that the quality of the water is in compliance with the specified global as well as regional standards. Organizations such as the World Health Organization (WHO) specify guidelines that are used to determine the specified use of the water [4].

In the mathematical modeling, numerous models to forecast and predict water quality are highly popular. Time series models like AR (auto-regressive), ARMA autoregressive–moving-average model), ARIMA (auto-regressive integrated moving average) and regression models, linear as well as multiple (MLR) are extremely popular and used extensively in the area of water quality prediction and monitoring [5]. With ever increasing data, both quality- and quantity-wise, the traditional mathematical models have not efficiently been able to satisfy the tests by researchers. With the evolution of computing capabilities and modern equipment, artificial neural network (ANN) models, and more data driven models have been developed by computational researchers. These models efficiently capture the fundamental relationships of water quality data from the previously available examples [6]. In addition, it has also been understood that ANNs require fewer assumptions and can attain higher precision limits in comparison to the traditional approaches [7]. Interestingly, ANN models are appropriate in solving the non-linear and uncertain problems due to their uncanny resemblance with the brain nervous system of a living organism. This supports the fact that researchers have moved from ARIMA and MLR to ANNs for water quality monitoring and forecasting [8].

In this chapter we will discuss the water quality parameters and standards set by some renowned global as well as regional organizations. Moving ahead we aim at projecting a vivid idea about the current prospects of ANNs explaining various architectures accompanied with some real-time examples of their application in water quality research.

5.2 Water quality parameters

An assessment of water quality needs evaluation of different parameters or variables which would then be compared to the minimum standards set for its designated use. This section presents some of the variables used in defining water quality.

Water quality parameters			
Physical variables	Chemical variables		Biological variables
Turbidity	pH	Copper and zinc	Bacteria
Temperature	Acidity	Hardness	Algae
Color	Alkalinity	Dissolved oxygen (DO)	Biological oxygen demand (BOD)
Taste and odor	Chloride	Chemical oxygen demand (COD)	Protozoa
Solids (TDS and TSS)	Chlorine residual	Toxic inorganic substances	Viruses
Electrical conductivity	Sulfate	Toxic organic Substances	
	Nitrogen	Radioactive substances	
	Fluoride	Iron and manganese	

5.2.1 Chemical variables

pH is one of the most important parameters of water quality. Furthermore, pH of water sample can be affected by other chemical variables such as presence of organic and inorganic toxicants. pH on the other hand is the regulating factor for determining the acidity as well as alkalinity of water samples. Although there are no standards set for acidity and alkalinity of water samples, higher values of either of them can drastically impact the potability of water [9]. Chloride has been observed to be present naturally in water obtained from lakes, streams and the ground, although public drinking water is suggested to have chloride concentrations below 250 mg L^{-1} [10]. In addition, further explanations about other parameters are present in much of the reviewed literature.

5.2.2 Physical variables

The physical variables of water quality include turbidity, odor, color, and taste. These parameters can be evidently observed and water with high turbidity, characteristic odor or taste and color are termed unfit for consumption. The solids within are determined experimentally by measuring the total dissolved solids (TDS) and total suspended solids (TSS) of water samples. Water containing below 900 mg L^{-1} has been considered acceptable by the WHO for consumption [11]. TDS may also be used to measure the ionic strength of water samples and also electrical conductivity. Typical conductivities as mentioned in literature for drinkable water ranges anything between 0.005 and 0.05 s m^{-1}. The relationship between TDS and electrical conductivity of water is as follows [10, 12]:

$$\mathrm{TDS}(\mathrm{mg\,L^{-1}}) \cong \mathrm{EC}(\mathrm{deciSiemens\,m^{-1}}) \times (0.55 - 0.7)$$

5.2.3 Biological variables

Biological variables are one of the most useful markers of water quality which can be identified either by the presence or lack of living organisms [10, 13]. Environmentalists survey aquatic life such as fish, insects and aquatic plants in the water bodies and a species diversity index (SDI) is calculated which is the basis for determination of its quality [13–15]; hence it has been the understanding that a healthy water body is that which has a large number of or a well-balanced diversity of species. There are also certain organisms, of which the presence or absence can be an indicator for the presence of certain contaminants in water based upon their tolerance to the substance in question [16].

5.3 Fundaments of machine learning

Patterns are prevalent in chemistry. Whether in the crystalline structures in solids, the branching of lipids, or the intricate sequences of functional groups, chemical patterns decide the fundamental properties of molecules and materials that are vital to address significant applications for society. Artificial intelligence (AI), and more specifically, machine learning (ML), has been utilized in recognizing and creating a vivid understanding of these patterns [17]. ML can be regarded as *a subgroup of artificial intelligence* which broke into the scene of computational researchers in the late 1970s. It focuses on extraction and automatically analyzing or manipulating the key information from large databases/datasets. The ML algorithms build a mathematical model which is based on sample data, this is also termed as *training data*, which is then utilized to make the projections and decisions without actually being explicitly programmed to perform the task again and again. According to Mitchell, learning can be defined as follows [18]:

> *A computer program is said to learn from experience E with respect to some class of tasks T and performance measure P if its performance at tasks in T, as measured by P, improves with experience E.*

The machine adopts new processes that were learned from previous iterations, familiarity, example, or analogy [19].

Learning can be *supervised* and *unsupervised.* In *supervised* learning algorithms are used to train the system using input data as well as correct output, whereas in *unsupervised* learning an algorithm identifies the framework of the dataset by dividing and then clustering the data [20]. Each of these learning types has certain models used for solving different problems. One of the most well-known ML models is ANNs, and this is the digital equivalent of human neurons and their functioning. They require few input variables and conditions; hence making the systems computationally inexpensive. ANNs are largely used in the solutions of forecast, predictions, and classification problems [21].

5.4 Artificial neural networks

A neural network can be defined as an interconnected network of simple signal processing units, '*neurons*'. A *neuron* is a human brain cell which processes all the human body signals coming through to the brain in the form of electrochemical impulses and responds with an appropriate sensory response. The signals are transported via *synaptic nodes* which join two different neurons [19] (figure 5.1).

The fundamental idea and motivation behind ANN is to mirror the characteristics of a human brain and its capacity to learn [22]. ANNs are similar to the human brain given that the algorithms can observe patterns in an adequate number of samples and then use this knowledge to simplify to other instances that have not yet been encountered. Despite a connection between the modeling strategies of ANNs and the human brain, ANNs do not mimic what the human brain does. As the biological neural networks are made up of neurons, the ANNs are contributed by *perceptrons*. A perceptron is simply the single functioning unit of an ANN. A mathematical model of a perceptron is analogous to that of a biological neuron [23] (figure 5.2).

A perceptron receives an input 'x_i' from the dendrites and an output is produced along its axon. The basic purpose of the perceptron is to add together the products of inputs 'x_i' and the weights 'w_i' associated with them, followed by comparing the summation to a certain threshold. The neuron will provide the output only when the threshold is surpassed. This threshold frequency can be represented as *activation function* denoted by f [22, 24].

ANNs are comprised of simple and highly interconnected perceptrons, that are like the human brain's neurons. Perceptrons are linked by weighted links to

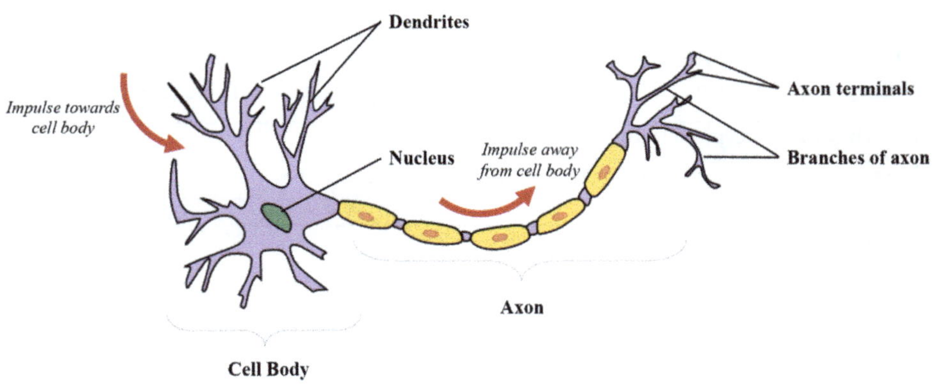

Figure 5.1. The human neuron. This neuron image has been adapted by the author from the Wikimedia website where it was made available under a CC BY-SA 3.0 licence. It is included within this chapter on that basis. It is attributed to the US National Cancer Institute's Surveillance, Epidemiology and End Results (SEER) Program.

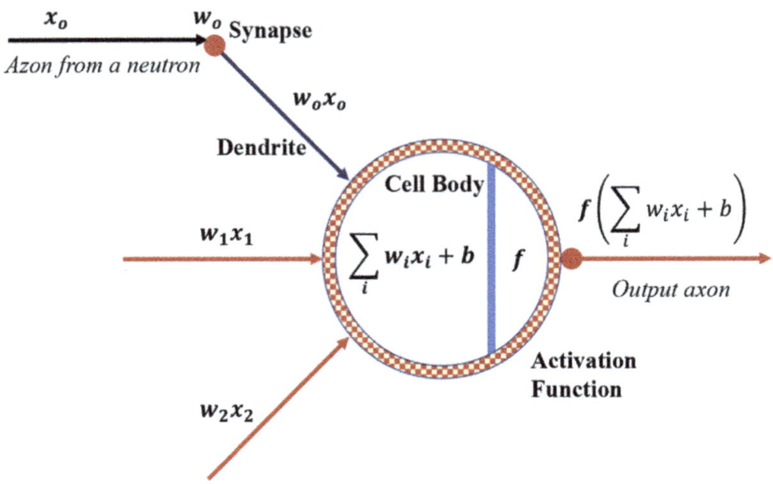

Figure 5.2. Schematic representation of a perceptron (analogous to a biological neuron) [23].

communicate information, with every perceptron receiving several inputs and producing just one output. An ANN mimics the human neural network's function in which neurons are represented through perceptrons, dendrites in accordance with perceptron inputs, axons as the perceptron outputs, and synapses by the weight of the input signals [25].

The mathematical operations related with a perceptron are:

$$y = \sum_{i=1}^{N} w_i x_i + b \tag{5.1}$$

$$z = f(y) \tag{5.2}$$

According to Agarwal, a perceptron learning algorithm is applied to a Euclidean instance space where $X \subseteq \mathbb{R}^d$. The algorithm then adapts the weights of the instances according to variations between predicted and input (real) values [26].

Algorithm 1 : Perceptron learning algorithm

input :*d-dimensional training set* X = ($x_1, x_2, x_3, \ldots, x_N$)

parameter :initial weight vector $\boldsymbol{w} \subseteq \mathbb{R}^d$

forinputs and weights: $\boldsymbol{i} \in (\mathbf{1, 2, 3, \ldots, N})$**do**

  $\hat{y} = f(\Sigma w_i x_i + b)$;

  receive the true value of y_i ;

  if$y_i \neq \hat{y}$ **then**

    $w_{i+1} \leftarrow w_i - \Delta w_i$:

Table 5.1. Various examples of ANNs based on the three basic architectures.

Architecture	Examples
Feedforward	Multilayer Perceptrons (MLPs), Radial Basis Function Neural Network (RBFNN), Graph Recurrent Neural Networks (GRNNs), Wavelet Neural Networks (WNNs), Cascade Correlation Neural Networks (CCNN), Modular Neural Networks (MNNs), Time Delay Neural Networks (TDNNs)
Recurrent	Recurrent Neural Networks (RNN), Long Short Term Memory networks (LSTM), Nonlinear autoregressive with exogenous input (NARX), Simple Recurrent Units (SRU), Time Lagged Recurrent Network (TLRN), Echo State Network (ESN), Ridge Regression Echo State Network (RESN)
Hybrid	LSTM-RNN, Fuzzy Wavelet NN (FNN-WNN), ARIMA-RBFNNs, ARIMA-ANN, Wavelet Artificial Neural Networks (WANN),

5.4.1 Architectures of ANN

The architecture of an ANN will be established by the manner in which the outputs of the perceptron are linked to the other perceptron. The neurons of a network are split into several groups, called layers. On the basis of these layers, single as well as multiple layer architectures can be created. Hence, an ANN comprises some single units, artificial neurons, linked with the individual coefficients called *processing elements* or *weights* 'w_i', that form the neural structure. The manner in which the neurons are connected to each other has a considerable impact on the function of the network. The architecture needs to be defined, which includes establishing the number of layers as well as setting the number of neurons per layer. This stage is followed by initialization of weight and updating the system based upon the training examples or initial inputs.

An ANN calculates the output by implementing a function to the summation of the values of the preceding layer and the values of the weights. Subsequently, the difference between the ANN's expected output and the actual value is calculated.

Although there are numerous architectures in ANNs, for the purpose of water quality monitoring and prediction, three model architectures have been vastly utilized (table 5.1). These are:

1. *Feedforward Architectures;*
2. *Recurrent Architectures;*
3. *Hybrid Architectures.*

5.4.1.1 Feedforward architectures

The expression '*feedforward*' implies that in the neurons, a connection exists from the neurons of the input layer to the subsequent neurons in the hidden layer, or from the hidden layer neurons to the output layer neurons. One other property of this architecture is that the neurons are not interconnected within the layer but the connection exists only from one layer to the other [28].

Feedforward network systems have the following characteristics [29]:

- Perceptrons are organized into layers, with the first layer getting in inputs and the last layer generating outputs.
- The intermediate layers have no relationship with the external world, and therefore are referred to as *hidden layers*.
- Each perceptron in a single layer is linked to every individual perceptron on the following layer. Consequently, the data is constantly 'feedforward' after one layer to the next, and that would explain why such networks are dubbed feedforward networks.

Learning in feedforward networks belongs in the domain of supervised learning, in which pairs of sample input and expected output datasets are fed into the network for numerous cycles, in order to ensure that the network 'learns' the correlation between the input and output (figure 5.3) [30]. MLPs are the most predominantly used algorithms in feedforward architectures. TDNNs is a subcategory of MLPs that learns chronological behavior from constant past and present signals [31]. The main difference between the RBFNNs and MLPs is that the hidden layer in the case of RBFNNs is self-coordinating, while for MLPs this is not the case, even though the structure of RBFNNs is comparable to that of MLPs. As a hub of RBF, the training or sample input weights can be specified by a clustering algorithm. For instance, the *k-means* algorithm is a frequently used weight defining algorithm [32].

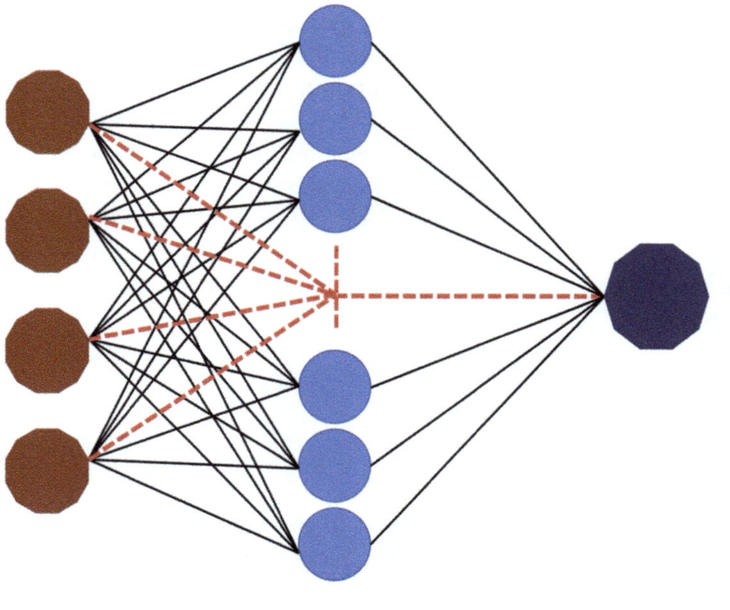

Figure 5.3. An example of ANN with input, hidden, and output layers [27].

5.4.1.2 Recurrent architectures

The Recurrent Neural Network (RNN) is the network that has loops, which enables the information to stay within network. RNN has a response and feedback connection embedded in the network, which facilitates the activations to stream back through into the loop, learn structures in addition to information to endure (figure 5.4). RNNs are very powerful in developing sequential data, speech or text as well as applied on non-sequential data to educate in a non-sequential manner [33].

In its the most common form, an RNN can be viewed as a weighted, directed, and cyclic chart that includes three unique types of nodes, such as the input, hidden, and output nodules [34]. While input and output layers are typified by feedforward connections, the hidden layers comprise recurrent units. Input nodes do not possess incoming connections, output nodes do not possess outgoing connections, whereas hidden nodes have both.

The Elman RNN is the most simple form of RNN (figure 5.4), also known as *Simple RNN* or *Vanilla RNN.* On the other hand, LSTM, Graded Recurring Unit (GRU) are rather more complex RNN architectures and can be understood as an extension to ERNNs. Echo State Networks (ESN) are somewhat different from the above mentioned RNNs, with three different layers—input, reservoir and readout layers. The characteristic feature of the reservoir layer is that it is at random and sporadically connected. The echo state property for which the internal states are especially reliant on the inputs is the most important factor to the ESN [35].

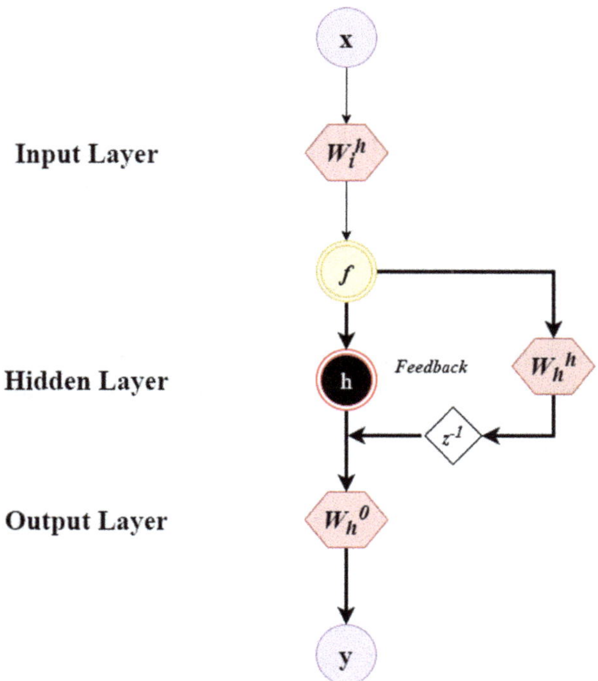

Figure 5.4. Schematic illustration of a simple RNN architecture. (Input x, output y, and hidden layers h represented by circles. The solid hexagons represent input, hidden, and output weights, respectively. Their values are generally set in the training.)

5.4.1.3 Hybrid architectures

Hybrid neural network systems are the computing systems which are primarily based on ANNs, although they also allow a symbolic rendition or communication with symbolic components. There is an increasing trend to use hybrid ANNs systems, which play an enormous role in modeling, for their ability to integrate with other traditional and more advanced modeling techniques. Hybrid prototypes of ANNs have been put forward, where various activation/transfer functions are designed for the nodes within the hidden layer. Various authors have suggested the hybridization of the various basis functions, either by using one single hybrid hidden layer or multiple connected pure layers [36].

Hybrid models are categorized into three different categories—*model-intensive, technique-intensive, and data-intensive* [37]. A few examples of hybrid architectures that are model-intensive systems are LSTM-RNN [38] and FNN-WNN [39]. On the other hand, the very essence of the technique intensive methods is to develop a modeling framework that has the ability to utilize the various technologies. Techniques that merge the ensemble methods or time series models that eliminate trends or periodicities such as the ARIMA-Radial Basis Function Neural Networks (ARIMA-RBFNNs) and ARIMA-ANN, which are a few examples of technique-intensive methods [40–42]. Wavelet analysis approaches, for example in WANN, have been able to offer some helpful information about the physical structure of the data-intensive models. In ANNs architectures, the estimate as well as the details element form the discrete wavelet transformation. Dimensionality reduction techniques like the principal component analysis (PCA) can decrease the dimension of the input data space in order to prevent duplication or redundancy [43, 44].

5.5 Real-time applications of ANN in water quality prediction

Several ML models have been implemented in order to determine and forecast the overall quality of water. They generally utilize certain forms of neural network. In this section we will discuss some real-time examples of various ANN models which have been utilized in water quality assessment. During the past decade, there has been a surge in ANN models being used for the purpose. Five out of six ML techniques employed to predict water quality make use of ANNs or several of their variants. Various other studies have also utilized back propagation neural networks (BPNNs) with different architectures [45–48]. Khan and See applied fuzzy neural networks FNNs and NARNNs, respectively [49]. These models are manipulated versions of ANNs.

Palani *et al* applied ANNs to forecast the weekly values of salinity, temperature, DO, and chlorophyll-a in East Johor Strait coastal water, Singapore. Data used for the above study have been collected during the period from December 1996 and June 1997 [50]. They pointed to the fact that the limited information and sample data was one of the shortcomings of their research and urged others to gather more data to recalibrate and revalidate the model. Dogan *et al* implemented ANN to predict the BOD, which is difficult to quantify and needs a minimum five days to get the definitive results at a waste water treatment plant (WWTP). Results indicated that COD remained the most effective variables for BOD estimation after carrying out the sensitivity assessment [51]. Miao *et al* applied BPNN for COD and ammonia nitrogen (NH_3-N) prediction in water samples. The entire datasets were normalized at the outset and subsequently divided into two sets with 80% as training and 20% as testing sets. The sigmoid transfer function that could establish the arbitrary nonlinear map between inputs and outputs was adopted [52]. Da Costa *et al* split the data into three unique sets using 50% as training, 25% for validation, and 25% of data as testing sets [53]. Shen *et al* utilized a golden section method to choose the hidden layer nodes in the BPNN [54]. Singh *et al* probed the partition method in assessing the relative significance of eleven environment variables to the resulting output layer. They split the datasets into the three aforementioned sets allocating 60%, 20%, and 20% as training, validation and testing sets, respectively. Results demonstrated that the expected values as predicted by the ANN model were in correspondence with the calculated values [55]. Yeon and co-workers further moved to combine precipitation (Precip) and discharge (Q) to achieve a one-step forecast of Q. Then, the linked system used the forecast value of Q and historical data of total organic carbon (TOC) to meet the needs for one-step prediction of TOC. Ultimately, the connected system had improved performance over individual ANN models [56].

Liu *et al* pointed to the fact that if more historical data were accessible, ANN models could have succeeded in providing better projections than a comparatively small dataset [7, 57]. Antanasijević *et al* examined the performance of RNN, GRNN, and MLP for small samples prediction. Results showed that the error of RNN in the test dataset was not more than 10%. Apart from that, the error obtained in GRNN was smaller than that of MLP [58]. Evrendilek and Karakaya removed the missing data directly and consequently employed discrete wavelet transforms (DWTs) with the orthogonal wavelet families for noise reduction in the data assessed by proximal sensors. The outcome suggested that the modeling impact on utilizing the Time Lagged Recurrent Network (TLRN) to the data post noise reduction proved to be superior to TLRN, time delay neural network (TDNN), and RNN [59]. Chang *et al* endeavored to utilize the NARX, a dynamic neural network, to model decade long seasonal water quality data. Subsequently, a 42-layer cross-validation algorithm was used to split the data. Results have shown that the NARX had outperformed BPNN due to the fact that it could capture the vital dynamic features of total phosphorous (TP) data [60]. Wang *et al* experiment with the prediction execution of LSTM, BPNN, Online sequential (OS)-ELM for DO, and TP. The results suggested that LSTM was more precise and able to be generalized than the aforementioned feedforward neural networks [61].

Yu *et al* applied five-fold cross-validation to segregate the data and implemented RBFNN to blend data from several sensors. The rate of convergence and the solution accuracy could be enhanced via a variant of PSO—integrated particle swarm optimization (IPSO). A comparison between the of prediction outcomes substantiated the efficacy of the hybrid model [62]. Zhao *et al* transformed the signal into an output linear system through the Kalman filter. The final outcome demonstrates that this hybrid method was a decent and efficient approach to water quality prediction [63]. Another research group led by Huang modeled the nonlinearity of data through a combination of the neural network, fuzzy logic, wavelet transform, and the genetic algorithms (GA). The results of these simulations demonstrated that this hybrid model can manage the problems of data instability or fluctuations [39]. Fijani and his team applied the variational mode decomposition (VMD) algorithm to decompose the maximum frequency component delivered by a complete ensemble empirical mode decomposition algorithm with adaptive noise (CEEMDAN) and applied ELM for modeling. Findings of their study pointed out that this hybrid model might be able to reduce error regardless if in root mean square or mean absolute error [64]. Jin *et al* suggested an enhancement modification variation, namely an improved genetic algorithm (IGA) to help avoid the position in which excellent individuals are cast off by the GA [65]. Li *et al* presented an evidence theory that possesses excellent data fusion capability, as it is able to reason with ambiguity to integrate the evidence from SRU, LSTM sources, Gated Recurrent Unit (GRU) in DO, pH, TP forecast, and ultimately attained a particular level of confidence. The enhanced probability assignment function of the evidence theory which was devised in accordance with the softmax function, can find a solution to the failure of weight allocation problems that are present in the conventional probability assignment function [7]. The application of this evidence theory can be further expanded as a general framework of uncertain reasoning. Tian and co-workers combined the ANN models that do not require a large quantity of training information with transfer learning (TL) due to the fact that the TL possesses the ability to transfer knowledge from earlier tasks to accurately forecast chlorophyll a with a dynamic range. The major difference between the TL and conventional ANNs techniques is the fact that the former does not necessarily have to learn each task from scratch whereas the latter does. Results showed that these hybrid models improved the generalization capacity contrasted with the dropout as well as the parameter norm penalties methods over the long-term application. At exactly the same time, the effect of mutable data distribution with regard to the models was decreased [66]. Yan *et al* employed mean value method by utilizing a median of *k* data prior and after to fix the wrong data and received the missing data in accordance with the values of model prediction of other water quality variables at the missing moment/position. The restricted condition of this particular model was that the data have been appropriately and normally distributed. Hence, it is uncertain whether the aforementioned method could be applied to some other prediction tasks that fail to meet the above conditions [67]. In another probe carried out by Yan *et al*, the group had suggested a hybrid optimized algorithm, such as PSO and GA, to augment BPNN with reasonable accuracy [68]. Liu and co-workers examined the DO projection, which took into account a temporal and spatial relationship. *Spatial relationship* is a reference to the spatial correlations between the external variables as opposed to the geographic allocations. This freshly intended attention-RNN model attained exceptional performance equally in short-term and long-term prediction [69].

5.6 Discussion

Numerous methods have been used in order to assess and predict the quality of water. This chapter is a comprehensive attempt to try and present a few of the

ML-based ANN models which have been used in predicting as well as assessment of water quality and its variables. Between the statistical, mathematical and ANN models, statistical models are fairly straightforward and easy to execute. Most of the studies that applied ML models used neural networks and applied them on DO or oxygen-related variables. ANNs offer an especially good option because they are computationally fast and necessitate a substantially smaller number of input parameters as well as the input conditions. ANNs, nevertheless, need a large pool of sample datasets for training. The experiments have considered data from various regions of the world. In water quality modeling, researchers tend to have a preference for cause-effect models over time series models. In developing nations like India, there is an issue of accessibility of data. In these circumstances, it is the hybrid models that are preferred if substantial parameter data is not available.

Data collection is an expensive and arduous process. This is primarily due to the costly equipment, restricted by the experimental time, and conditions. The ANN model represents a data-driven model, so acquiring sufficient data constitutes the basis of modeling. The need to collect as much data as possible is ever important for modeling ANNs. To tackle this need, scientists need to take into account two factors. The first is, whether or not the historical knowledge of the output variable may be gathered and the latter being, the data-selection of strategy by the researchers when the historical evidence of output variables are tough to measure. If researchers can gather historical observations of target variables, they can process the data and model it. This also brings us to the fact that precision and accuracy along with the effectiveness of ANNs in water quality assessments ride on three basic planning steps, as shown in figure 5.5.

In recent times there has been an emergence of a different kind of ANNs which are not discussed in section 5.4. Convolutional Neural Networks or CNNs is a type of feedforward architecture which has been historically used in imaging fields. They have discrete layers which comprise input, convolution, pooling, full connection, and output layers. Over the past few years, CNN has already been used as an evolving method in water quality prediction. The utilization of convolution can be realized more than once to expose the relationship between the parameters concealed in the input matrix [70]. Nevertheless, since the objective of the prediction model is to isolate the potential factors instead of simply bringing in the convolutional layer's results to an even higher level, the pooling layer will be removed. Meanwhile, the number of calculations can be reduced. Deep Belief Network (DBN) is a type of neural network that is based upon deep learning that is like the feedforward framework and has been used extensively during the past few years. However, the research studies regarding a dynamic way of determining the structure are rarely investigated. To defeat these limitations a self-organizing deep belief networks (SODBN) has been suggested. The framework of the SODBN is not decided by the artificial experience but the automatic growing and pruning algorithm (AGP) [71, 72]. Particularly, the concealed layers and neurons are altered by the AGP at the beginning. Subsequently, the weights of the SODBN are continually adjusted through the process of self-organization. Ultimately, several aspects of network performance, like running time and prediction accuracy have been enhanced. Over the last few years,

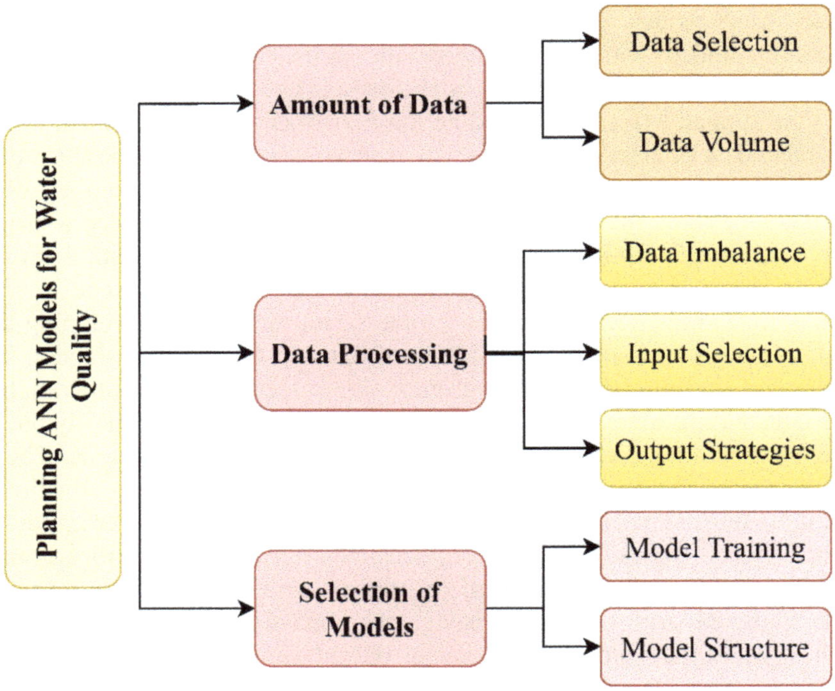

Figure 5.5. Planning of ANN models for water quality assessment and prediction.

CNN, as a brand-new feedforward neural network approach, has been used for water quality prediction. However, its implementation is rather limited. RNN has excellent memory ability, so it will be able to make full use of the historical data and provide a solid foundation for achieving long-term forecast of water quality parameters. Hybrid models must be further developed since they are not a replacement for conventional technologies, rather they are a blend of their advantages. Scientists may also refer to the collaborative methodologies, TL technology, and evidence theory in order to recalibrate their models to improve accuracy, generalization ability and adapt to prediction uncertainties

References

[1] US Environmental Protection Agency 2009 *Technical Assistance Document for Reporting of Daily Air Quality—Air Quality Index* (Washington, DC: US Environmental Protection Agency)

[2] World Economic Forum 2020 The Global Risks Report 2020

[3] Wang Y *et al* 2017 Water quality prediction method based on LSTM neural network *12th Int. Conf. on Intelligent Systems and Knowledge Engineering (ISKE)* (Piscataway, NJ: IEEE)

[4] World Health Organization 2017 *Guidelines for Drinking-Water Quality: Fourth Edition Incorporating First Addendum* 4th edn + 1st add edn (Geneva: World Health Organization)

[5] Araghinejad S 2013 *Data-Driven Modeling: Using MATLAB® in Water Resources and Environmental Engineering* vol 67 (Berlin: Springer Science & Business Media)

[6] Zhang G, Patuwo B E and Hu M Y 1998 Forecasting with artificial neural networks: the state of the art *Int. J. Forecast.* **14** 35–62
[7] Li L *et al* 2019 Water quality prediction based on recurrent neural network and improved evidence theory: a case study of Qiantang River, China *Environ. Sci. Pollut. Res.* **26** 19879–96
[8] Zhang L, Zou Z and Shan W 2017 Development of a method for comprehensive water quality forecasting and its application in Miyun reservoir of Beijing, China *J. Environ. Sci.* **56** 240–46
[9] Flanagan P J 1988 *Parameters of Water Quality: Interpretation and Standards* (Dublin: Environmental Research Unit)
[10] Water Environment Federation and American Water Works Association 2005 *Standard Methods for the Examination of Water and Wastewater* (Washington, DC: American Public Health Association (APHA))
[11] World Health Organization 1996 Total dissolved solids in drinking-water *Guidelines for Drinking-Water Quality* vol 2 2nd edn (Geneva: World Health Organization)
[12] Metcalf *et al* 2003 *Wastewater Engineering: Treatment and Reuse* (New York: McGraw-Hill)
[13] Cole S *et al* 2000 *Guidelines for managing water quality impacts within UK European marine sites* UK Marine SACs Project, Peterborough, UK
[14] Alabaster J S and Lloyd R S 2013 *Water Quality Criteria for Freshwater Fish* (Amsterdam: Elsevier)
[15] Cariappa M P 2004 Basic environmental technology—water supply, waste management and pollution control *Med. J. Armed Forces India* **60** 206
[16] Abbas S H *et al* 2014 Biosorption of heavy metals: a review *J. Chem. Sci. Technol.* **3** 74–102
[17] Cova T F G G and Pais A A C C 2019 Deep learning for deep chemistry: optimizing the prediction of chemical patterns *Front. Chem.* **7** 809
[18] Mitchell T 1997 Introduction to machine learning *Mach. Learn.* **7** 2–5
[19] Negnevitsky M 2005 *Artificial Intelligence: A Guide to Intelligent Systems* (London: Pearson Education)
[20] Ahmed U *et al* 2019 Efficient water quality prediction using supervised machine learning *Water* **11** 11
[21] Harzallah S *et al* 2018 Eddy current modelling using multi-layer perceptron neural networks for detecting surface cracks *Frattura Integr. Strutt.* **12** 147–55
[22] Fiesler E and Beale R 2020 *Handbook of Neural Computation* (Boca Raton, FL: CRC Press)
[23] Minsky M and Papert S A 2017 *Perceptrons: An Introduction to Computational Geometry* (Cambridge, MA: MIT Press)
[24] Fiesler E and Beale R 1997 *Handbook of Neural Computation: Supplement 1* (Boca Raton, FL: CRC Press)
[25] Hasson U, Nastase S A and Goldstein A 2020 Direct fit to nature: an evolutionary perspective on biological and artificial neural networks *Neuron* **105** 416–34
[26] Agarwal S 2019 *Lecture Notes Machine Learning*, *Online Learning* (and *Perceptron*)
[27] Kumar S 2018 Prediction of Jatropha-algae biodiesel blend oil yield with the application of artificial neural networks technique *Energy Sources* A **41** 1285–95
[28] Nourani V, Alami M T and Vousoughi F D 2016 Self-organizing map clustering technique for ANN-based spatiotemporal modeling of groundwater quality parameters *J. Hydroinf.* **18** 288–309
[29] Skapura D M 1996 *Building Neural Networks* (Reading, MA: Addison-Wesley Professional)

[30] Mehrotra K and Mohan C 1997 *Rakka. Elements of Artificial Neural Networks* (Cambridge, MA: MIT Press)

[31] Gholamreza A *et al* 2016 Application of artificial neural networks to predict total dissolved solids in the river Zayanderud, Iran *Environ. Eng. Res.* **21** 333–40

[32] Csábrági A *et al* 2017 Application of artificial neural networks to the forecasting of dissolved oxygen content in the Hungarian section of the river Danube *Ecol. Eng.* **100** 63–72

[33] Sak H, Senior A W and Beaufays F 2014 Long short-term memory recurrent neural network architectures for large scale acoustic modeling *15th Annual Conf. of the Int. Speech Communication Association (Singapore)*

[34] Zhang S *et al* 2016 *Architectural Complexity Measures of Recurrent Neural NetworksArchitectural Complexity Measures* arXiv:1602.08210

[35] Zhao J *et al* 2018 Water quality prediction in the waste water treatment process based on ridge regression echo state network *IOP Conf. Ser.: Mater. Sci. Eng.* **435** 012025

[36] Yan J *et al* 2019 Application of a hybrid optimized BP network model to estimate water quality parameters of Beihai Lake in Beijing *Appl. Sci.* **9** 1863

[37] Maier H R *et al* 2010 Methods used for the development of neural networks for the prediction of water resource variables in river systems: current status and future directions *Environ. Model. Softw.* **25** 891–909

[38] Ye Q *et al* 2019 River water quality parameters prediction method based on LSTM-RNN model *2019 Chinese Control And Decision Conf. (CCDC), Nanchang, China* (Piscataway, NJ: IEEE)

[39] Huang M *et al* 2017 A new efficient hybrid intelligent model for biodegradation process of DMP with fuzzy wavelet neural networks *Sci. Rep.* **7** 41239

[40] Elkiran G, Nourani V and Abba S I 2019 Multi-step ahead modelling of river water quality parameters using ensemble artificial intelligence-based approach *J. Hydrol.* **577** 123962

[41] Deng W *et al* 2014 Water quality prediction based on a novel hybrid model of ARIMA and RBF neural network *2014 IEEE 3rd Int. Conf. on Cloud Computing and Intelligence Systems* (Piscataway, NJ: IEEE)

[42] Faruk D Ö 2010 A hybrid neural network and ARIMA model for water quality time series prediction *Eng. Appl. Artif. Intell.* **23** 586–94

[43] Sakizadeh M, Malian A and Ahmadpour E 2016 Groundwater quality modeling with a small data set *Groundwater* **54** 115–20

[44] Ravansalar M and Rajaee T 2015 Evaluation of wavelet performance via an ANN-based electrical conductivity prediction model *Environ. Monit. Assess.* **187** 366

[45] Palani S, Liong S-Y and Tkalich P 2008 An ANN application for water quality forecasting *Mar. Pollut. Bull.* **56** 1586–97

[46] Muhammad S Y *et al* 2015 Classification model for water quality using machine learning techniques *Int. J. Softw. Eng. Appl.* **9** 45–52

[47] Chine W-H *et al* 2009 Artificial neural networks for water quality prediction in a reservoir *2009 Second Int. Workshop on Computer Science and Engineering* (Piscataway, NJ: IEEE)

[48] Zhu C and Hao Z 2009 Fuzzy neural network model and its application in water quality evaluation *2009 Int. Conf. on Environmental Science and Information Application Technology* (Piscataway, NJ: IEEE)

[49] Khan Y and See C S 2016 Predicting and analyzing water quality using machine learning: a comprehensive model *2016 IEEE Long Island Systems, Applications and Technology Conf. (LISAT)* (Piscataway, NJ: IEEE)

[50] Chuah A L 1998 Water Quality Study of the East Johor Strait *Master's Thesis* National University of Singapore

[51] Dogan E *et al* 2008 Application of artificial neural networks to estimate wastewater treatment plant inlet biochemical oxygen demand *Environ. Prog.* **27** 439–46

[52] Miao Q *et al* 2009 Water quality prediction of Moshui river in China based on BP neural network *2009 Int. Conf. on Computational Intelligence and Natural Computing* (Piscataway, NJ: IEEE)

[53] da Costa A O S *et al* 2009 Use of neural networks for monitoring surface water quality changes in a neotropical urban stream *Environ. Monit. Assess.* **155** 527–38

[54] Shen X, Chen M and Yu J 2009 Water environment monitoring system based on neural networks for shrimp cultivation *2009 Int. Conf. on Artificial Intelligence and Computational Intelligence* (Piscataway, NJ: IEEE)

[55] Singh K P *et al* 2009 Artificial neural network modeling of the river water quality—a case study *Ecol. Model.* **220** 888–95

[56] Yeon I, Jun K and Lee H J 2009 The improvement of total organic carbon forecasting using neural networks discharge model *Environ. Technol.* **30** 45–51

[57] Liu S *et al* 2012 Prediction of dissolved oxygen content in aquaculture of hyriopsis Cumingii using Elman neural network *Computer and Computing Technologies in Agriculture V* (Berlin: Springer)

[58] Antanasijević D *et al* 2013 Modelling of dissolved oxygen content using artificial neural networks: Danube River, North Serbia, case study *Environ. Sci. Pollut. Res.* **20** 9006–13

[59] Evrendilek F and Karakaya N 2014 Monitoring diel dissolved oxygen dynamics through integrating wavelet denoising and temporal neural networks *Environ. Monit. Assess.* **186** 1583–91

[60] Chang F-J *et al* 2016 Estimating spatio-temporal dynamics of stream total phosphate concentration by soft computing techniques *Sci. Total Environ.* **562** 228–36

[61] Wang Y *et al* 2017 Water quality prediction method based on LSTM neural network *Intelligent Systems and Knowledge Engineering (ISKE)* (Piscataway, NJ: IEEE)

[62] Yu H *et al* 2016 Dissolved oxygen content prediction in crab culture using a hybrid intelligent method *Sci. Rep.* **6** 27292

[63] Zhao Y, Zou Z and Wang S 2015 A back propagation neural network model based on Kalman filter for water quality prediction *11th Int. Conf. on Natural Computation (ICNC)* (Piscataway, NJ: IEEE)

[64] Fijani E *et al* 2019 Design and implementation of a hybrid model based on two-layer decomposition method coupled with extreme learning machines to support real-time environmental monitoring of water quality parameters *Sci. Total Environ.* **648** 839–53

[65] Jin T *et al* 2019 A data-driven model for real-time water quality prediction and early warning by an integration method *Environ. Sci. Pollut. Res.* **26** 30374–85

[66] Tian W, Liao Z and Wang X 2019 Transfer learning for neural network model in chlorophyll-a dynamics prediction *Environ. Sci. Pollut. Res.* **26** 29857–71

[67] Yan J *et al* 2019 Application of a parallel particle swarm optimization-long short term memory model to improve water quality data *Water* **11** 1317

[68] Yan J *et al* 2019 Application of a hybrid optimized BP network model to estimate water quality parameters of Beihai Lake in Beijing *Appl. Sci.* **9** 1863

[69] Liu Y *et al* 2019 Attention-based recurrent neural networks for accurate short-term and long-term dissolved oxygen prediction *Comput. Electron. Agric.* **165** 104964

[70] Ta X and Wei Y 2018 Research on a dissolved oxygen prediction method for recirculating aquaculture systems based on a convolution neural network *Comput. Electron. Agric.* **145** 302–10

[71] Qiao J *et al* 2018 A self-organizing deep belief network for nonlinear system modeling *Appl. Soft Comput.* **65** 170–83

[72] Lin Q *et al* 2018 Deep-learning based approach for forecast of water quality in intensive shrimp ponds *Indian J. Fish* **65** 75–80

www.ingramcontent.com/pod-product-compliance
Ingram Content Group UK Ltd.
Pitfield, Milton Keynes, MK11 3LW, UK
UKHW051849100926
8556IPUK00005B/537